Mathematics
Education
Innovation

Tom Button

Series editor
Catherine Berry
Series advisor
Roger Porkess

A LEVEL
FURTHER
MATHEMATICS

Further Pure Maths with Technology

HODDER
EDUCATION
AN HACHETTE UK COMPANY

Hachette UK's policy is to use papers that are natural, renewable and recyclable products and made from wood grown in sustainable forests. The logging and manufacturing processes are expected to conform to the environmental regulations of the country of origin.

Orders: please contact Bookpoint Ltd, 130 Park Drive, Milton Park, Abingdon, Oxon OX14 4SE. Telephone: (44) 01235 827720. Fax: (44) 01235 400401. Email education@bookpoint.co.uk Lines are open from 9 a.m. to 5 p.m., Monday to Saturday, with a 24-hour message answering service. You can also order through our website: www.hoddereducation.co.uk

ISBN: 978 1 5104 0359 8

© Tom Button and MEI 2018

First published in 2018 by

Hodder Education,
An Hachette UK Company
Carmelite House
50 Victoria Embankment
London EC4Y 0DZ

www.hoddereducation.co.uk

Impression number 10 9 8 7 6 5 4 3 2 1

Year 2022 2021 2020 2019 2018

Cover photo © Zoonar GmbH / Alamy Stock Photo

Illustrations by Integra Software Services Pvt. Ltd., Pondicherry, India

Typeset in Bembo Std, 11/13 by Integra Software Services Pvt. Ltd., Pondicherry, India

Printed in U.K.

A catalogue record for this title is available from the British Library.

Contents

Getting the most from this book

Mathematics is not only a beautiful and exciting subject in its own right but also one that underpins many other branches of learning. It is consequently fundamental to our national wellbeing.

This textbook covers the content of Y436 Further Pure with Technology, one of the minor options in the MEI A Level Further Mathematics specification. Some students will begin this course in year 12 alongside the A level course, whereas others only begin Further Mathematics when they have completed the full A Level Mathematics and so will have already met some of the topics, or background to topics, covered in *MEI A Level Mathematics (Year 2)*. This book has been written with all these users in mind.

The book begins with exploring the ideas relating to the shapes of curves expressed in cartesian, polar or parametric form including how they can plotted using graphing software and analysed with a Computer Algebra System (CAS). The next chapter "explores" differential equations by plotting *tangent fields* in graphing software and also analysing their solutions through either the use of a CAS, where they can be solved analytically, or a spreadsheet, to solve them numerically. The final chapter introduces some ideas in number theory and how a programming language can be used to search for solutions to problems.

Between 2014 and 2016 A Level Mathematics and Further Mathematics were very substantially revised, for first teaching in 2017. Changes include increased emphasis on

- Problem solving
- Mathematical rigour
- Use of technology
- Modelling.

This book embraces these ideas. A large number of exercise questions involve elements of problem solving and require rigorous logical argument. Technology is used throughout the book and the number theory chapter includes examples of how mathematical problems can be modelled in software.

When studying the content of this book you will need access to a graph-plotter, a spreadsheet and a Computer Algebra System. You are expected to be confident in using your software and this book suggests plenty of activities to support you in practising this. You will also need access to a programming language: Python has been used throughout this book. Specific guidance for programming in Python is included in this book.

Throughout the book, the emphasis is on understanding and interpretation rather than mere routine calculations, but the various exercises do nonetheless provide plenty of scope for practising techniques. In addition, extensive online support, including further questions, is available by subscription to MEI's Integral website, http://integralmaths.org.

This book can be used alongside the study of A level Mathematics; however some knowledge of parametric equations of curves (covered in A level Mathematics) and polar equations of curves (covered in A level Further Mathematics Pure Core) is required in chapter 1. In chapter 2, knowledge of first order differential equations (covered in A level Mathematics) is required.

At the end of each chapter there is a list of key points covered as well as a summary of the new knowledge (learning outcomes) that readers should have gained.

Two common features of the book are Activities and Discussion points. These serve rather different purposes. The Activities are designed to help readers get into the thought processes of the new work that they are about to meet. The Discussion points invite readers to talk about particular points with their fellow students and their teacher and so enhance their understanding.

Answers to all exercise questions are provided at the back of the book, and also online at www.hoddereducation.co.uk/MEIFurtherMathsFurtherPurewithTechnology

This is a 4th edition MEI textbook so some of the material is well tried and tested. However, as a consequence of the changes to A Level requirements in Further Mathematics, some parts of the book are either new material or have been very substantially rewritten.

Catherine Berry
Roger Porkess

1

Investigation of curves

Many mathematical relationships can be represented geometrically as curves, and these curves can often give insights into those relationships. In this topic, you will use graphing software and a computer algebra system (CAS) to investigate curves. You will learn to look for and recognise important properties of curves, or families of curves. You are expected to be able to make conjectures about these properties and then verify them using analytical techniques.

You will meet a variety of curves expressed as cartesian, parametric or polar equations. You are not expected to know the particular properties of specific curves, or families of curves. Instead, you should be able to select and apply the skills learned in this topic to investigate them.

Equations of curves

In this topic, the equation of a curve will be expressed in one of the following ways:

- A **cartesian equation** expresses a direct relationship between the x and y coordinates of points on the curve. These can be expressed *explicitly*, where an expression for y is given in terms of x, e.g. $y = \dfrac{x^3 + x^2 - 6x}{x^2 - 2x - 3}$. They can also be expressed *implicitly* as an equation linking x and y that does not have y as the subject, e.g. $x^2 + y^3 + y = 1$.

- A **polar equation** expresses the distance r from the pole (i.e. the origin) against the angle θ, usually given in radians, measured anticlockwise from the positive horizontal axis, e.g. $r = 1 + 2\sin\dfrac{\theta}{3}$. Some conventions/software show sections of the curve that correspond to negative values of r as a broken line and others display this as a solid line.

- **Parametric equations** are pairs of expressions giving the x and y coordinates in terms of a parameter t, e.g. $x = 3\cos\dfrac{t}{2}$, $y = \sin 2t, 0 \leq t < 4\pi$. The limits of the parameter are also often stated. There can be non-identical parametric equations that give the same curve.

In a parametric equation, the parameter does not necessarily represent a geometrical quantity such as an angle. For this reason, it is preferable to use t instead of θ as the parameter to avoid this confusion.

> **Note**
>
> Polar coordinates are covered in *MEI A Level Further Mathematics Core Year 2*.

> **Note**
>
> Parametric equations are covered in *MEI A level Mathematics Year 2*.

1 Equations and properties of curves

Use of software: Graphing

You will need access to graphing software that can plot curves in cartesian, polar and parametric forms. Your software should also be able to plot families of curves, with sliders (or equivalent) for parameters.

The graphs in Figure 1.1 have been plotted using graphing software.

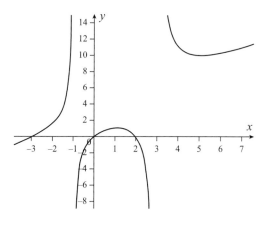

$$y = \frac{x^3 + x^2 - 6x}{x^2 - 2x - 3}$$

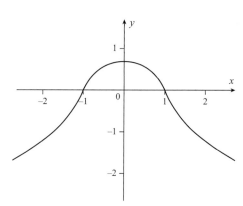

$$x^2 + y^3 + y = 1$$

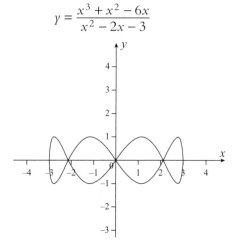

$$x = 3\cos\frac{t}{2}, \; y = \sin 2t, 0 \le t < 4\pi$$

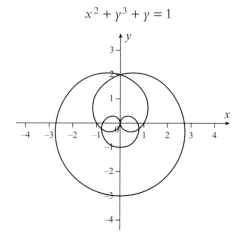

$$r = 1 + 2\sin\frac{\theta}{3}$$

Figure 1.1

ACTIVITY 1.1

Use your graphing software to plot the curves in Figure 1.1. Try plotting some similar curves.

Your graphing software should allow you to enter a parameter into the equation of a curve and view its effect on the family of curves when it is changed. This is usually represented by a slider whereby the value can be changed and the curve updates automatically.

Example 1.1

A family of curves has cartesian equation $y = \dfrac{x^2 + kx - 2}{x - 1}$.

Sketch the curve for the cases $k = -1$, $k = 1$ and $k = 3$.

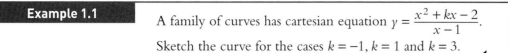

First plot the curve in your software so that you can vary the value of k. Then observe that it has one of three shapes.

Solution

$k = -1$

For $k < 1$, the curve has two distinct branches either side of an asymptote at $x = 1$, both of which are increasing (i.e. the derivative is positive for all values).

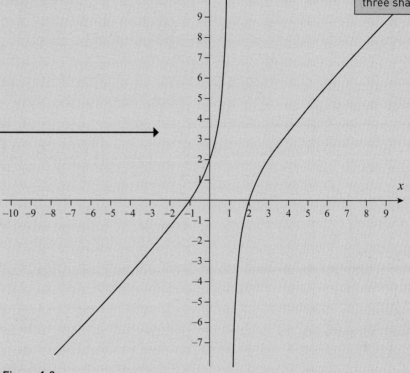

Figure 1.2

$k = 1$

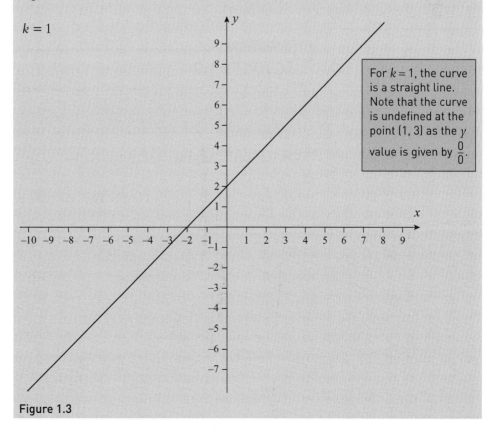

For $k = 1$, the curve is a straight line. Note that the curve is undefined at the point $(1, 3)$ as the y value is given by $\dfrac{0}{0}$.

Figure 1.3

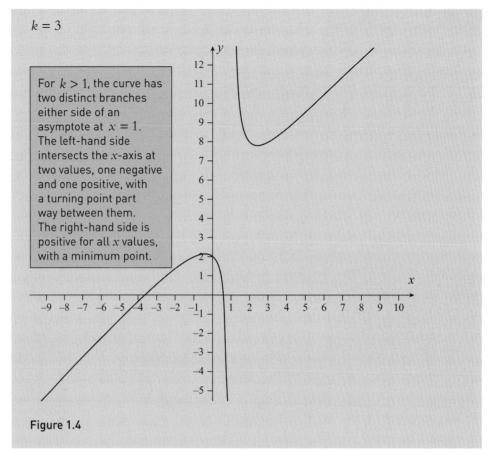

$k = 3$

For $k > 1$, the curve has two distinct branches either side of an asymptote at $x = 1$. The left-hand side intersects the x-axis at two values, one negative and one positive, with a turning point part way between them. The right-hand side is positive for all x values, with a minimum point.

Figure 1.4

Discussion points

➜ How does changing the parameter affect the following families of curves?

$y = 3x + c$

$y = x^2 + bx + 4$

$y = e^{kx}$

➜ Give some further examples of equations of curves with parameters in them where varying the parameter results in:

● a translation

● a stretch

● a reflection.

You should note the distinction between the use of the words 'plot' and 'sketch'. When you **plot** a curve you accurately mark the points on graph paper, or enter the equation into a graph plotter to create a plot of the curve. A **sketch** shows the shape of the curve and any important points, such as if you copied the plot from the graph plotter onto paper.

ACTIVITY 1.2

Find some examples of equations of curves in cartesian, parametric and polar form that you have met in A Level Maths/Further Maths and explore the effect on the graphs of replacing one, or more, of the numbers in the equation with a parameter.

Use of software: CAS

You will need access to a computer algebra system (CAS) that can solve equations, differentiate, integrate and find limits. The emphasis in this topic is on selecting and applying techniques. You can use CAS, when appropriate, to perform algebraic processes: you might know various non-technology based methods for some of these processes, but using CAS can be less time-consuming.

Your software should be able to:

● find the solution to an equation, e.g.

$$x^4 + 5x^3 - \left(2 + \frac{a}{4}\right)x^2 - \frac{5a}{4}x + \frac{a}{2} = 0 \Rightarrow x = \frac{-\sqrt{33} - 5}{2}, \frac{\sqrt{33} - 5}{2}, \frac{-\sqrt{a}}{2}, \frac{\sqrt{a}}{2}$$

● find derivatives, e.g. $y = \dfrac{\sqrt{x} \sin kx}{x + 1} \Rightarrow \dfrac{\mathrm{d}y}{\mathrm{d}x} = \dfrac{\sin kx}{2\sqrt{x}\,(x + 1)} - \sqrt{x}\,\dfrac{\sin kx}{(x + 1)^2} + \dfrac{k\sqrt{x}\cos x}{x + 1}$

- find integrals, e.g. $\int x^2 \sin(x+k)\,\mathrm{d}x = 2x\sin(x+k)+(2-x^2)\cos(x+k)+c$

- find limits, e.g. $\displaystyle\lim_{x\to 5}\frac{x^2-2x-15}{x^2-6x+5}=2$.

ACTIVITY 1.3

Use your software to verify the previous examples. Find some further examples of equations, derivatives and integrals that you have met in A Level Maths/ Further Maths and explore using your CAS to solve them.
Note that all software has limitations and, particularly when using CAS, there are often cases where you have to interpret the output or do some further mathematical analysis.

Features of curves

When investigating curves you are expected to be able to identify certain features.

Specific features you should be looking for include:

- important points such as intersections with the axes and stationary points

- any reflectional or rotational symmetry

- whether the curve is bounded, either in terms of being limited to a particular range of x or y values or limited to being within a certain distance from the pole

- any asymptotes: asymptotes can be vertical, horizontal or oblique; a curve can also approach another curve asymptotically

- the existence of any cusps or loops: a **cusp** is a point where two branches of a curve meet, with the branches having a common tangent at the point; a **loop** is a continuous section of a curve whose end points meet.

Full coverage of asymptotes and cusps is given on pages 11–13.

Example 1.2

A family of curves has parametric equations $x = \dfrac{t^2}{1+t^2}, y = t^3 + at$. Sketch the curves for the cases where $a = -1$, $a = 0$ and $a = 1$. State any common features of these three curves and any features that are distinct to one of the three cases.

You should plot the curves in your software with a variable parameter a. Explore other values of a to help you to get a sense of how the family of curves are related to each other.

Solution

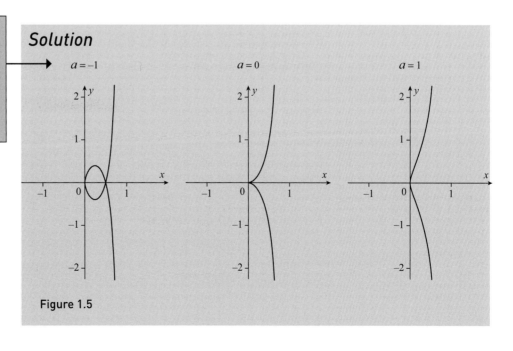

Figure 1.5

All of these curves have the following features in common:

- They pass through the point $(0,0)$.
- They are symmetrical about the x-axis.
- They are bounded by the region $0 \leq x < 1$.
- They have an asymptote at $x = 1$.

Features that can be observed in one of the three cases are:

- The case $a = -1$ has a loop.
- The case $a = -1$ intersects the positive x-axis.
- The case $a = 0$ has a cusp at $(0,0)$.

Discussion point

→ In Example 1.2, the features have been conjectured based on observing the curves. How could you show analytically that the curves have these features?

You should make conjectures about the features of curves by observing them and then use analytical techniques to verify mathematically that the features are as observed. You should be able to identify whether these features remain fixed or vary as the parameter in a family of curves is varied.

Converting between forms

It is often useful to be able to convert the equation of a curve from one form to another, as certain features will be easier to show in one form rather than the other.

Cartesian to polar

To convert between a cartesian equation and a polar equation for a curve, substitute $x = r\cos\theta$ and $y = r\sin\theta$ into the expression linking x and y and then rearrange to make r the subject.

For example, $x + 2y = 10 \Rightarrow r\cos\theta + 2r\sin\theta = 10 \Rightarrow r = \dfrac{10}{\cos\theta + 2\sin\theta}$.

Polar to cartesian

You can convert from polar to cartesian form directly by substituting $r^2 = x^2 + y^2$ into $r^2 = (f(\theta))^2$. You might find it useful to use $\tan\theta = \dfrac{y}{x}$. You should also look out for any instances of $r\cos\theta$ or $r\sin\theta$ that can be replaced with x and y, respectively.

For example, $r = 3\cos\theta \Rightarrow r^2 = 3r\cos\theta \Rightarrow x^2 + y^2 = 3x$.

Polar to parametric

Given $r = f(\theta)$, substitute $r = f(t)$ into $x = r\cos t$ and $y = r\sin t$.

For example, $r = \tan\theta \Rightarrow x = \tan t\cos t$ and $y = \tan t\sin t$.
This can be simplified to $x = \sin t, y = \tan t\sin t$.

Parametric to cartesian

Attempt to eliminate the parameter to obtain an expression solely in terms of x and y.

For example, $x = 2t + 1, y = t^2 \Rightarrow y = \left(\dfrac{x-1}{2}\right)^2$.

① Sketch the following curves on separate axes for the cases shown. State any common features of all cases or features that are distinct to one of the cases.

    (i) $y = \dfrac{x^2 - k}{x + k}$ for the cases $k = -1$, $k = 0.5$ and $k = 2$.

    (ii) $x = a\cos t + \cos at$, $y = a\sin t - \sin at$, $0 \le t < 2\pi$ for the cases $a = 2$, $a = 3$ and $a = 4$.

    (iii) $r = a + \cos\theta$ for the cases $a = 0.5$, $a = 1$ and $a = 2$.

② Complete the following.

    (i)   Show that the cartesian equation

$$\big((x + 1)^2 + y^2\big)\big((x - 1)^2 + y^2\big) = 1$$

    can be written in polar form as

$$r^2 = 2\cos 2\theta.$$

    (ii)   Convert the polar equation $r = 2a\sin\theta\tan\theta$ into cartesian form.

    (iii)   Convert the polar equation $r = 2a\sin\theta\tan\theta$ into parametric form.

    (iv)   Convert the parametric equations $x = at$, $y = \dfrac{a}{1 + t}$ into cartesian form.

2 Derivatives of curves

Gradients of tangents to curves

You can use differentiation to find an expression for the gradient of a tangent to a curve at a general point $\dfrac{dy}{dx}$ for a curve expressed in cartesian form. An expression containing $\dfrac{dy}{dx}$ can be found where the curve is expressed either explicitly, such as $y = \dfrac{x^3 + x^2 - 6x}{x^2 - 2x - 3}$, or implicitly, such as $x^2 + y^3 + y = 1$.

You can also use differentiation to find an expression for the gradient of a tangent to a curve at a general point $\dfrac{dy}{dx}$ for a curve expressed parametrically. This can be found by using

$$\frac{dy}{dx} = \frac{\dfrac{dy}{dt}}{\dfrac{dx}{dt}}.$$

By convention, the derivative of a parametric curve is expressed in terms of the parameter t.

> **Note**
> --------------------
> Differentiation of cartesian and parametric equations is covered in A Level Mathematics. In this topic, it is assumed that you will be able to find such derivatives with CAS.

Example 1.3

Sketch the curve given by the parametric equations $x = \sin 2t$, $y = \sin t$, $0 \leq t < 2\pi$ and find the points on the curve where the tangent is parallel to the y-axis.

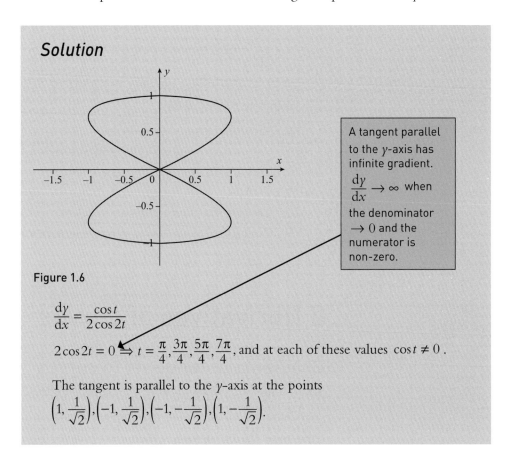

Solution

Figure 1.6

A tangent parallel to the y-axis has infinite gradient. $\dfrac{dy}{dx} \to \infty$ when the denominator $\to 0$ and the numerator is non-zero.

$$\frac{dy}{dx} = \frac{\cos t}{2\cos 2t}$$

$2\cos 2t = 0 \Rightarrow t = \dfrac{\pi}{4}, \dfrac{3\pi}{4}, \dfrac{5\pi}{4}, \dfrac{7\pi}{4}$, and at each of these values $\cos t \neq 0$.

The tangent is parallel to the y-axis at the points

$$\left(1, \frac{1}{\sqrt{2}}\right), \left(-1, \frac{1}{\sqrt{2}}\right), \left(-1, -\frac{1}{\sqrt{2}}\right), \left(1, -\frac{1}{\sqrt{2}}\right).$$

Discussion point

→ What geometrical property of a polar curve is represented by $\dfrac{dr}{d\theta}$?

Discussion point

The curve with parametric equation $x = \cos 2t$, $y = \sin t$, $0 \leq t < 2\pi$ has derivative $\dfrac{dy}{dx} = \dfrac{\cos t}{-2\sin 2t}$.

→ Does the curve have a tangent that is parallel to the y-axis at all the points where $-2\sin 2t = 0$?

When working with a polar curve you can use differentiation to find an expression for $\dfrac{dr}{d\theta}$; however, this does not represent the gradient of the tangent to a curve at a point. You can find the gradient of the tangent to a polar curve by finding an expression for $\dfrac{dy}{dx}$ when the equation of the curve is expressed parametrically as $x = r\cos\theta$ and $y = r\sin\theta$:

$$\frac{dy}{dx} = \frac{\dfrac{dr}{d\theta}\sin\theta + r\cos\theta}{\dfrac{dr}{d\theta}\cos\theta - r\sin\theta}.$$

Example 1.4

(i) Sketch the curve with the polar equation $r = 1 + \cos\theta$.

(ii) Find the points on the curve where the tangent is parallel to the x-axis.

Solution

(i)

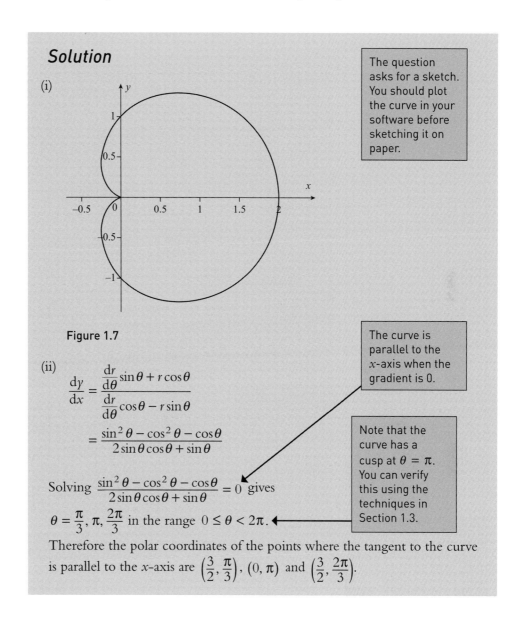

> The question asks for a sketch. You should plot the curve in your software before sketching it on paper.

Figure 1.7

(ii)
$$\frac{dy}{dx} = \frac{\dfrac{dr}{d\theta}\sin\theta + r\cos\theta}{\dfrac{dr}{d\theta}\cos\theta - r\sin\theta}$$

$$= \frac{\sin^2\theta - \cos^2\theta - \cos\theta}{2\sin\theta\cos\theta + \sin\theta}$$

> The curve is parallel to the x-axis when the gradient is 0.

> Note that the curve has a cusp at $\theta = \pi$. You can verify this using the techniques in Section 1.3.

Solving $\dfrac{\sin^2\theta - \cos^2\theta - \cos\theta}{2\sin\theta\cos\theta + \sin\theta} = 0$ gives

$\theta = \dfrac{\pi}{3}, \pi, \dfrac{2\pi}{3}$ in the range $0 \le \theta < 2\pi$.

Therefore the polar coordinates of the points where the tangent to the curve is parallel to the x-axis are $\left(\dfrac{3}{2}, \dfrac{\pi}{3}\right)$, $(0, \pi)$ and $\left(\dfrac{3}{2}, \dfrac{2\pi}{3}\right)$.

Equations of chords, tangents and normals

You should be able to find the equation of a chord between two points on a curve or the equation of a tangent or a normal to a curve at a point. For tangents and normals, you should find it useful to express them in the form $y - y_1 = m(x - x_1)$.

Example 1.5

(i) Sketch the curve given by the parametric equations $x = t - \sin t$, $y = 1 - \cos t$, $0 \leq t < 2\pi$.

(ii) Show that the normal to the curve passes through the point $(t, 0)$.

Solution

(i)

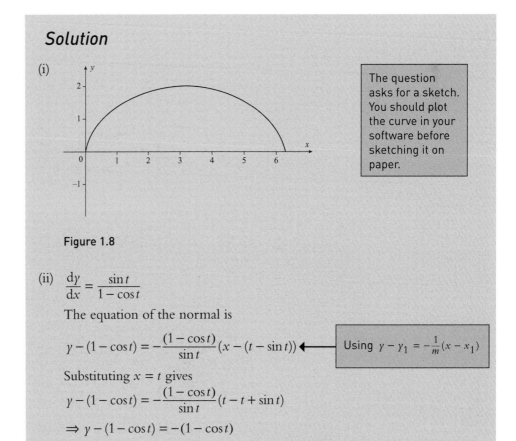

The question asks for a sketch. You should plot the curve in your software before sketching it on paper.

Figure 1.8

(ii) $\dfrac{dy}{dx} = \dfrac{\sin t}{1 - \cos t}$

The equation of the normal is

$y - (1 - \cos t) = -\dfrac{(1 - \cos t)}{\sin t}(x - (t - \sin t)) \longleftarrow$ Using $y - y_1 = -\dfrac{1}{m}(x - x_1)$

Substituting $x = t$ gives

$y - (1 - \cos t) = -\dfrac{(1 - \cos t)}{\sin t}(t - t + \sin t)$

$\Rightarrow y - (1 - \cos t) = -(1 - \cos t)$

$\Rightarrow y = 0$, i.e. the point $(t, 0)$ lies on the normal to the curve.

Exercise 1.2

① A curve has equation $x = \dfrac{t^2}{1 + t^2}$, $y = t^3 + t$.

(i) Sketch the curve.

(ii) Show that the tangent to the curve is never parallel to the x-axis and there is only one point on the curve where the tangent is parallel to the y-axis.

② Show that, for the curve with equation $y = \dfrac{1}{x}$, the tangent to the curve at a point does not cross the curve at any other points.

3 Limiting behaviour

Asymptotes

The curve with equation $y = \dfrac{3x - 1}{x - 2}$, whose graph is shown in Figure 1.9, has a vertical asymptote and a horizontal asymptote.

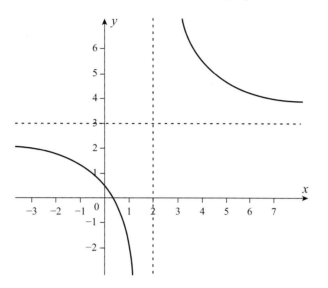

Figure 1.9

You can use limits to show that the curve has asymptotes at $x = 2$ and $y = 3$.

The denominator $x - 2$ means that $y \to \infty$ as $x \to 2$. This means that the curve has a vertical asymptote at $x = 2$.

The equation of the curve can be written as $y = 3 + \dfrac{5}{x - 2}$, and since $\dfrac{5}{x - 2} \to 0$ as $x \to \infty$, then $y \to 3$. This can be written as $\lim\limits_{x \to \infty} \dfrac{3x - 1}{x - 2} = 3$ and shows that the curve has a horizontal asymptote at $y = 3$.

In general, if a function can be written in the form $f(x) = c + \dfrac{p(x)}{q(x)}$, where $p(x)$ and $q(x)$ are polynomials and the degree of $p(x)$ is less than the degree of $q(x)$, then the graph of $y = f(x)$ will have a horizontal asymptote at $y = c$.

This idea can be extended to curves that have oblique asymptotes as well as to cases where curves are approached asymptotically.

An **oblique asymptote** is defined as a line with non-zero gradient that the curve approaches as $x \to \infty$. The graph of the equation $y = f(x)$ has an oblique asymptote of $y = mx + c$ if the function $f(x)$ can be written in the form $f(x) = mx + c + \dfrac{p(x)}{q(x)}$, where $p(x)$ and $q(x)$ are polynomials and $p(x)$ has degree less than the degree of $q(x)$.

For example, each member of the family of curves with equations $y = \dfrac{x^2 + kx + x + k + 2}{x + 1}$ has a vertical asymptote at $x = -1$. The equations can be written $y = x + k + \dfrac{2}{x + 1}$, and therefore each curve has an oblique asymptote at $y = x + k$. The curve for the case $k = 2$ is shown in Figure 1.10.

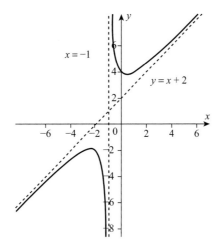

Figure 1.10

If the function f(x) can be written in the form $f(x) = r(x) + \dfrac{p(x)}{q(x)}$, where p($x$) and q($x$) are polynomials and p($x$) has degree less than the degree of q(x), then you say that the graph of $y = f(x)$ approaches the graph of $y = r(x)$ asymptotically.

For example, the curve $y = \dfrac{x^3 + 1}{x}$ has a vertical asymptote at $x = 0$. The equation can be written $y = x^2 + \dfrac{1}{x}$, and therefore the curve approaches $y = x^2$ as x tends to infinity. The curve is shown in Figure 1.11.

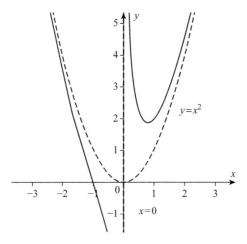

Figure 1.11

Cusps

A **cusp** is a point on a curve where two branches meet and the branches share a common tangent. To show that a point is a cusp, you need to show that the point is defined on the curve and that the gradient of the tangent to the curve approaches the same limit as the point is reached from either branch.

The full definition of a branch of a curve is beyond the scope of this book, but you should be able to observe branches as distinct parts of curves based on the plots from your software.

Example 1.6

Show that the curve $x = t^2$, $y = t^3 + 2$ has a cusp at the point $(0, 2)$.

Solution

A plot of the curve shows that there are two branches either side of $(0, 2)$.

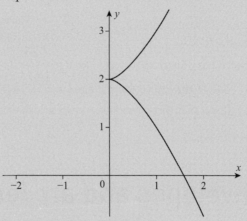

Figure 1.12

The derivative of the curve is

$$\frac{\mathrm{d}y}{\mathrm{d}x} = \frac{\mathrm{d}y/\mathrm{d}t}{\mathrm{d}x/\mathrm{d}t}$$

$$= \frac{3t^2}{2t}$$

$$= \frac{3}{2}t$$

> Your software should have the facility to evaluate limits approached from above and below using functions such as $\lim\limits_{t \to 0^+}$ and $\lim\limits_{t \to 0^-}$.

Substituting $t = 0$ into the parametric equations gives $x = 0$ and $y = 2$; therefore the point is defined.

The limit of the derivative as the curve is approached from either side can then be evaluated.

Limit approached from below: $\lim\limits_{t \to 0^-} \left(\frac{3}{2}t\right) = 0$.

Limit approached from above: $\lim\limits_{t \to 0^+} \left(\frac{3}{2}t\right) = 0$.

As the gradient of the tangent to the curve approaches the same limit from either side, the point is a cusp.

Exercise 1.3

① A curve has equation $y = \dfrac{2 + kx - x^3}{x^2 - k}$.

    (i)   Sketch the curve for the cases $k = -1$, $k = 0$ and $k = 4$.

    (ii)   State the equations of the asymptotes when:

       (a) $k > 0$

       (b) $k = 0$

       (c) $k < 0$.

② A curve has equation $x = t^3 + kt, y = \dfrac{t^2}{1 + t^2}$.

    (i)   Sketch the curve for the cases $k = -1$, $k = 0$ and $k = 1$.

    (ii)   Show that the curve has a cusp for the case $k = 0$.

4 Envelopes and arc lengths

Envelopes

You might already be familiar with the locus definition of a parabola: the set of points such that the distance from a fixed point (the focus) is equal to the perpendicular distance from a fixed line (the directrix).

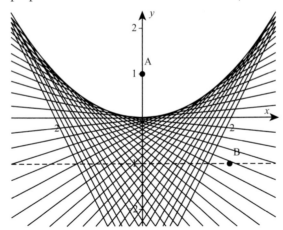

Figure 1.13

This definition can be used to generate a set of straight lines that show the parabola. In Figure 1.13, the point A = (0, 1), the line $x = -1$ and a point B lying on the line have been drawn. The perpendicular bisectors of AB for different positions of B are shown. You can generate your own version of this diagram using geometry software if you have access to it.

The family of lines generated are all tangential to a curve, and it is this curve that is 'seen' in the figure. The curve that every curve in a family touches is known as the **envelope** of the family.

An equation for the envelope can be derived by considering points on the lines close to each other in the family. To do this, you first need to be able to express the family of curves, using a parameter such as p, in the form $f(x, y, p) = 0$.

In Example 1.6, the family of lines can be generated using the equation $px - 2y - \dfrac{p^2}{2} = 0$ for different values of p.

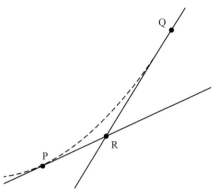

Figure 1.14

Two of the lines are shown in Figure 1.14 for the values p and $p + \delta p$. These lines touch the envelope at P and Q, respectively, and cross each other at R. This diagram suggests that as Q moves closer to P then R moves closer to P, i.e. P is the limiting position as $\delta p \to 0$.

The coordinates of R satisfy both $f(x, y, p) = 0$ and $f(x, y, p + \delta p) = 0$, and so they also satisfy $f(x, y, p) - f(x, y, p + \delta p) = 0$ and therefore $\dfrac{f(x, y, p) - f(x, y, p + \delta p)}{\delta p} = 0$. Therefore the coordinates of P (the limiting position as $\delta p) \to 0$) satisfy $\displaystyle\lim_{\delta p \to 0} \dfrac{f(x, y, p + \delta p) - f(x, y, p)}{\delta p} = 0$. This is written as $\dfrac{\partial}{\partial p} f(x, y, p) = 0$ and is known as the **partial derivative** of f. Partial derivatives are covered in full in the book *MEI Extra Pure*; for this topic, it is sufficient to know that partial differentiation involves differentiating with respect to one variable while treating the other variables as constants.

For the previous example where $f(x, y, p) = px - 2y - \dfrac{p^2}{2}$,

> Your CAS should be able to differentiate with respect to p.

$$\dfrac{\partial}{\partial p} f(x, y, p) = x - p.$$

Solve $px - 2y - \dfrac{p^2}{2} = 0$ and $x - p = 0$ simultaneously to eliminate p.

Substituting $p = x$ into $px - 2y - \dfrac{p^2}{2} = 0$ gives

$$x^2 - 2y - \dfrac{x^2}{2} = 0 \Rightarrow y = \dfrac{x^2}{4}.$$

> If you have used geometry software to generate the family of curves above, you can check that the envelope is $y = \dfrac{x^2}{4}$.

Example 1.7

Find the envelope of the family of circles that pass through the origin and have their centres on the parabola $y = x^2 + 1$.

Solution

The points on $y = x^2 + 1$ have coordinates $(p, p^2 + 1)$ for different values of the parameter p. The equations of the circles are given by

$$(x - p)^2 + (y - (p^2 + 1))^2 = p^2 + (p^2 + 1)^2$$

$$\Rightarrow x^2 + y^2 - 2y - 2px - 2p^2y = 0.$$

$$\frac{\partial}{\partial p} f(x, y, p) = -2x - 4py$$

> Remember that you can use CAS for expanding, differentiating with respect to p and substituting into equations.

$$\frac{\partial}{\partial p} f(x, y, p) = 0 \Rightarrow p = -\frac{x}{2y} \text{ or } x = 0, y = 0.$$

Substituting $p = -\dfrac{x}{2y}$ into $x^2 + y^2 - 2y - 2px - 2p^2y = 0$ gives

$\dfrac{2y^3 + 2x^2y + x^2 - 4y^2}{2y} = 0$. Multiplying by $2y$ (provided $y \neq 0$) gives

$2y^3 + 2x^2y + x^2 - 4y^2 = 0$. As $x = 0, y = 0$ also satisfies this equation, a cartesian equation of the full envelope can be given as

$2y^3 + 2x^2y + x^2 - 4y^2 = 0.$

Figure 1.15 shows the curve $y = x^2 + 1$ as a dotted line, the circles as dashed lines and the envelope as a solid line.

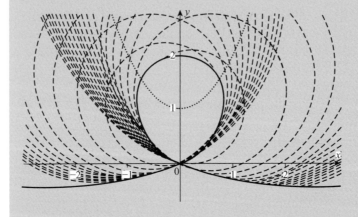

Figure 1.15

Arc lengths

The length of an arc on a curve can be found by considering chords joining points on the arc. Figure 1.16 shows two different sets of chords on the same arc. In finding the length of the arc from P_1 to P_5, the sum of the four chords $P_1P_2 + P_2P_3 + P_3P_4 + P_4P_5$ is a better approximation than the sum of the two chords $P_1P_3 + P_3P_5$.

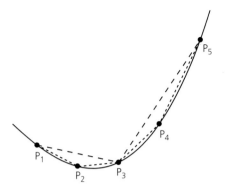

Figure 1.16

An assumption is being made that, for two points P and Q on a curve, the length of the chord PQ tends to the length of the arc PQ as Q gets closer to P, i.e. $\dfrac{\text{arc PQ}}{\text{chord PQ}} \to 1$ as $Q \to P$. This assumption holds for any curves that you will meet in this book; however, there are some fractal curves where this assumption does not hold, but these are beyond the scope of this topic.

Finding arc lengths on curves expressed as parametric equations

To find the length of an arc s on a curve given by parametric equations for x and y in terms of t, consider a pair of points P and Q close to each other with coordinates (x, y) and $(x + \delta x, y + \delta y)$, corresponding to parametric values t and δt, respectively. The length of the arc PQ is δs. This is shown in Figure 1.17.

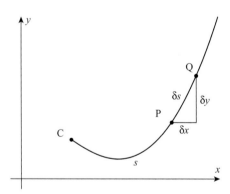

Figure 1.17

It is important that the function for the positive square root is used, as lengths of arcs are strictly positive.

The chord length PQ is $|PQ| = \sqrt{(\delta x)^2 + (\delta y)^2}$.

Therefore $\dfrac{\delta s}{\delta t} = \dfrac{\delta s}{|PQ|} \times \dfrac{|PQ|}{\delta t}$ can be rewritten as $\dfrac{\delta s}{\delta t} = \dfrac{\delta s}{|PQ|} \times \sqrt{\left(\dfrac{\delta x}{\delta t}\right)^2 + \left(\dfrac{\delta y}{\delta t}\right)^2}$,

As $\delta t \to 0$, $\dfrac{\delta s}{\delta t}, \dfrac{\delta x}{\delta t}, \dfrac{\delta y}{\delta t}$ tend to $\dfrac{ds}{dt}, \dfrac{dx}{dt}, \dfrac{dy}{dt}$, respectively, and, by the assumption stated earlier, $\dfrac{\delta s}{|PQ|} \to 1$. So taking limits as $\delta t \to 0$ gives the result

$$\frac{ds}{dt} = \sqrt{\left(\frac{dx}{dt}\right)^2 + \left(\frac{dy}{dt}\right)^2}$$

and hence the length of the arc s can be found using

$$s = \int \sqrt{\left(\frac{dx}{dt}\right)^2 + \left(\frac{dy}{dt}\right)^2} \, dt.$$

Example 1.8

Sketch the curve with parametric equations $x = a\cos^3 t$, $y = a\sin^3 t$, $0 \le t < 2\pi$, where a is a positive constant, and find the length of this curve.

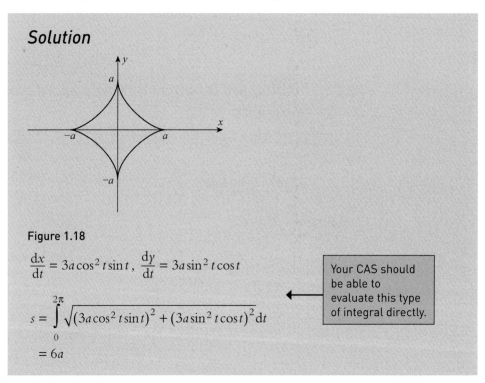

Solution

Figure 1.18

$$\frac{dx}{dt} = 3a\cos^2 t \sin t, \quad \frac{dy}{dt} = 3a\sin^2 t \cos t$$

Your CAS should be able to evaluate this type of integral directly.

$$s = \int_0^{2\pi} \sqrt{(3a\cos^2 t \sin t)^2 + (3a\sin^2 t \cos t)^2} \, dt$$

$$= 6a$$

Finding arc lengths on curves expressed as cartesian equations

For curves expressed as cartesian equations, a similar method can be used, by considering the equation as a parametric equation where $x = t$. This gives $\dfrac{dx}{dt} = 1$, and hence the arc length is given by

$$s = \int \sqrt{1 + \left(\frac{dy}{dx}\right)^2} \, dx.$$

Finding arc lengths on curves expressed as polar equations

For curves expressed as polar equations, you can find the arc length using $x = r\cos\theta$ and $y = r\sin\theta$ and considering the parameter as θ.

$$\frac{dx}{d\theta} = \frac{dr}{d\theta}\cos\theta - r\sin\theta \text{ and } \frac{dy}{d\theta} = \frac{dr}{d\theta}\sin\theta + r\cos\theta.$$

$$\left(\frac{dx}{d\theta}\right)^2 = \left(\frac{dr}{d\theta}\right)^2\cos^2\theta - 2r\left(\frac{dr}{d\theta}\right)\cos\theta\sin\theta + r^2\sin^2\theta$$

$$\left(\frac{dy}{d\theta}\right)^2 = \left(\frac{dr}{d\theta}\right)^2\sin^2\theta + 2r\left(\frac{dr}{d\theta}\right)\cos\theta\sin\theta + r^2\cos^2\theta$$

So $\left(\frac{dx}{d\theta}\right)^2 + \left(\frac{dy}{d\theta}\right)^2 = \left(\frac{dr}{d\theta}\right)^2 + r^2$ and hence the arc length is given by

$$s = \int\sqrt{\left(\frac{dr}{d\theta}\right)^2 + r^2}\,d\theta.$$

Example 1.9

Sketch the curve $r = \cos\theta + 1$ and find the length of this curve.

Solution

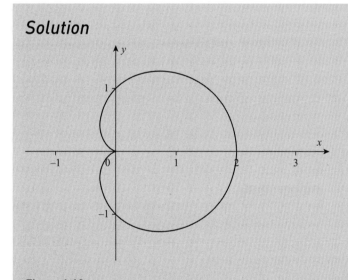

Figure 1.19

$$\frac{dr}{d\theta} = -\sin\theta$$

$$s = \int_0^{2\pi}\sqrt{(-\sin\theta)^2 + (\cos\theta + 1)^2}\,d\theta$$

$$= 8$$

Exercise 1.4

① Find the equation of the envelope of the family of circles that pass through the origin and have their centres on the curve with equation $y = x^2$.

② Find, by integrating, the lengths of the following curves.

  (i) $y = x^2$ from $x = 0$ to $x = a$.

  (ii) $x = \dfrac{1}{1+t^2}$, $y = \dfrac{t}{1+t^2}$ from $t = 0$ to $t = 1$.

  (iii) The cardioid $r = a(1 + \cos\theta)$.

LEARNING OUTCOMES

Now you have finished this chapter, you should be able to:

➤ plot a family of curves in graphing software, in cartesian, polar or parametric forms, and be able to use a slider to vary any parameters

➤ use CAS to solve equations, find or evaluate derivatives/integrals and find limits

➤ use the vocabulary associated with curves, such as asymptote, cusp, loop, bounded and terms relating to symmetry

➤ find, describe and generalise the properties of curves

➤ convert equations between cartesian and polar forms

➤ convert equations from polar to parametric form

➤ convert equations from parametric to cartesian form

➤ find the gradient of the tangent to a curve at a point

➤ work with equations of tangents, chords and normals

➤ calculate arc length by integrating for a curve expressed in cartesian, parametric or polar form

➤ find the envelope of a family of curves.

KEY POINTS

1 When investigating families of curves, some of the features to look for are: important points such as intersections with the axes and stationary points; any reflectional or rotational symmetry; whether the curve is bounded; any asymptotes; the existence of any cusps or loops.

2 To convert an equation from cartesian form to polar form, substitute $x = r\cos\theta$ and $y = r\sin\theta$ into the expression linking x and y and then rearrange to make r the subject.

3 To convert from polar to cartesian form directly, substitute $r^2 = x^2 + y^2$ into $r^2 = (f(\theta))^2$. Alternatively, replace any instances of $\tan\theta$ with $\dfrac{y}{x}$, $r\cos\theta$ with x or $y\sin\theta$ with y.

4 To convert from polar to parametric form, substitute $r = f(t)$ into $x = r\cos t$ and $y = r\sin t$.

5 To convert from parametric to cartesian form, eliminate the parameter to obtain an expression solely in terms of x and y.

6   The gradient of the tangent to a curve expressed parametrically at a general

point $\dfrac{dy}{dx}$ for a curve is $\dfrac{dy}{dx} = \dfrac{\dfrac{dy}{dt}}{\dfrac{dx}{dt}}$.

7   The gradient of the tangent to a curve expressed as a polar equation is

$$\dfrac{dy}{dx} = \dfrac{\dfrac{dr}{d\theta}\sin\theta + r\cos\theta}{\dfrac{dr}{d\theta}\cos\theta - r\sin\theta}.$$

8   Vertical asymptotes for curves can be found by looking for values of x that will give a value of 0 in the denominator of any fractions in the equation for the curve.

9   Horizontal and oblique asymptotes for curves can be found by evaluating the equation for the curve as $x \to \infty$. This method can also be used for curves that are approached asymptotically.

10 A cusp is a point where two branches of a curve meet, and the tangent to the curve tends to the same gradient as the point is approached along either branch.

11 The envelope of a family of curves can be found by writing the equation of the curve in the form $f(x, y, p) = 0$, using a parameter such as p, and then solving simultaneously the equations $f(x, y, p) = 0$ and $\dfrac{\partial}{\partial p}f(x, y, p) = 0$.

12 The length of an arc s on a curve can be found using either $s = \displaystyle\int \sqrt{1 + \left(\dfrac{dy}{dx}\right)^2}\, dx$,

$s = \displaystyle\int \sqrt{\left(\dfrac{dx}{dt}\right)^2 + \left(\dfrac{dy}{dt}\right)^2}\, dt$ or $s = \displaystyle\int \sqrt{\left(\dfrac{dr}{d\theta}\right)^2 + r^2}\, d\theta$, depending on whether the curve is expressed in cartesian, parametric or polar form.

2 Exploring differential equations

In mathematics the study of quantities that change continuously is called **calculus**. The rate of change of a quantity is called its **derivative**, and equations that involve the derivative are called **differential equations**.

At its simplest, a differential equation provides information about the rate of change of one variable with respect to another in terms of those variables, for example,

$$\frac{dy}{dx} = x + y \quad \text{or} \quad \frac{dP}{dt} = 0.2P.$$

You will study only first order differential equations in this topic. **A first order differential equation** features only first derivatives such as $\frac{dy}{dx}$, but no higher derivatives such as $\frac{d^2y}{dx^2}$.

The **solution** to a differential equation gives a relationship between the variables themselves and does not involve derivatives. When solving differential equations you will work with both general and particular solutions.

A **general solution** to a first order differential equation $\frac{dy}{dx} = f(x, y)$ is an expression relating x and y that involves a constant of integration, although this is not necessarily just a '$+c$'.

A **particular solution** to a first order differential equation $\frac{dy}{dx} = f(x, y)$ is an expression relating x and y that does not involve a constant of integration. This is usually found by substituting a value of x and the corresponding value of y into the general solution.

For example, the differential equation $\frac{dy}{dx} = 3xy^2$ has a general solution $y = \frac{2}{2c - 3x^2}$. The particular solution that passes through the point $(1, 2)$ is $y = \frac{2}{4 - 3x^2}$.

Various methods for solving differential equations using integration are covered in *MEI A Level Mathematics Year 2* and *MEI A Level Further Mathematics For Core Year 2*. In this topic, you will explore first order differential equations using technology. You will see how using a graph plotter to produce tangent fields can give you insights into the behaviour of solutions to differential equations. You will also explore the exact solutions to differential equations that can be solved analytically using a computer algebra system (CAS). For differential equations that cannot be solved analytically, you will learn some numerical techniques for approximating solutions that can be implemented on a spreadsheet.

1 Tangent fields

Plotting tangent fields

A differential equation gives you information about the rate of change of the dependent variable, but not about the actual value of the variable at any instant. The information about the rate of change can be represented graphically as a **tangent field** (sometimes called a direction field or a slope field), and this can be used to sketch approximate solution curves for the differential equation.

For example, the differential equation $\dfrac{dy}{dx} = 2x$ tells you that the gradient of the tangent to the curve of any particular solution is $2x$. Therefore, particular solutions that pass through $(3, 1), (3, -5)$ or $(3, 8.31)$ have tangents of gradient 6 at these points. Similarly, any particular solution has a tangent with gradient -4 at the point on the curve where $x = -2$.

You can represent this on a graph of y against x by drawing direction indicators over a selection of points to create a tangent field for the differential equation. A **direction indicator** is a short dash with the appropriate gradient at that point. Your software should have the facility to plot a tangent field for a differential equation $\dfrac{dy}{dx} = f(x, y)$.

A tangent field for $\dfrac{dy}{dx} = 2x$ is shown in Figure 2.1. The general solution to this differential equation is $y = x^2 + c$. The particular solutions have been drawn for $c = -3, c = 0$ and $c = 2$ in Figure 2.2.

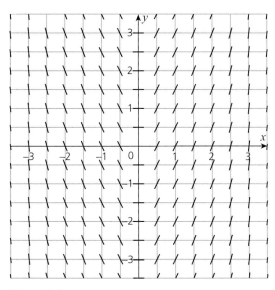

Figure 2.1

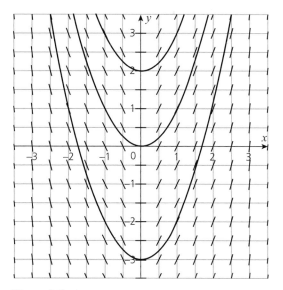

Figure 2.2

When plotting tangent fields the software often chooses the points at which to draw the direction indicators. When you make a sketch of the tangent field based on the computer output, you should choose sufficient points to ensure that the general shape of the tangent field is conveyed but not so many as to make the task of sketching it too time-consuming.

Interpreting tangent fields

Plotting tangent fields can allow you to make observations about the solutions to a differential equation without necessarily solving it. The tangent field for the differential equation $\dfrac{dy}{dx} = k - y$ for the case $k = 2$ is shown in Figure 2.3.

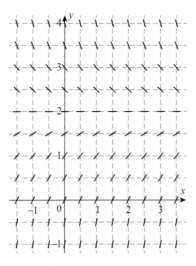

Figure 2.3

From this you can conjecture that all the particular solutions in this case have an asymptote at $y = 2$. Your software should allow you to vary the value of k and observe the effect on the tangent fields generated.

In the case of $\dfrac{dy}{dx} = k - y$, it is possible to conjecture that:

- any particular solution has an asymptote at $y = k$
- particular solutions that pass through points where $y > k$ are decreasing functions, and particular solutions that pass through points where $y < k$ are increasing functions.

When interpreting tangent fields, it is often useful to look at the curves along which the direction indicators have the same gradient. These curves are called the **isoclines**.

The tangent field for the differential equation $\dfrac{dy}{dx} = 2x$, shown in Figure 2.1, has isoclines at each x value. For example, all the points that lie on the line $x = -1$ have direction indicators with gradient $x = -2$.

The tangent field for the differential equation $\dfrac{dy}{dx} = y^2 - x$, shown in Figure 2.4, has isoclines at $y^2 - x = m$. Two isoclines are shown: $x = y^2 - 2$ and $x = y^2 + 1$.

Discussion point
→ Can an isocline ever be a particular solution?

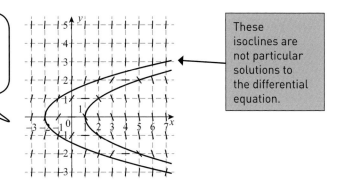

These isoclines are not particular solutions to the differential equation.

Figure 2.4

ACTIVITY 2.1

Find some further examples of differential equations and plot their tangent fields. Explore the effect on the graphs of replacing one, or more, of the numbers in these differential equations with a parameter.

Exercise 2.1

① The differential equation $\dfrac{\mathrm{d}y}{\mathrm{d}x} = \dfrac{a - xy}{x^2}$ is defined for $x > 0$ and $y > 0$.

  (i) Sketch the tangent field for the cases:

    (a) $a = 1$

    (b) $a = 4$.

  (ii) For what values of y, expressed in terms of a and x, is the particular solution that passes through the point (x, y) a decreasing function at that point?

② For the differential equation $\dfrac{\mathrm{d}y}{\mathrm{d}x} = x^2 + y^2 - 2y - 3$,

  (i) sketch the tangent field

  (ii) describe the shape of any isoclines.

2 Analytical solutions of differential equations

Finding solutions

Some differential equations, such as $\dfrac{\mathrm{d}y}{\mathrm{d}x} = 2x$, $\dfrac{\mathrm{d}p}{\mathrm{d}t} = \dfrac{5p}{t(2 + 3t)}$ and $x\dfrac{\mathrm{d}y}{\mathrm{d}x} + 5y = \dfrac{2}{x}$, can

be solved analytically. For differential equations such as these it is possible to give the general solution to the differential equation in the form $y = \mathrm{f}(x)$, where x and y are the independent and dependent variables, respectively. Where a differential equation has an analytical solution, you can use CAS to obtain both general and particular solutions directly.

Some differential equations cannot be solved analytically, and these will be covered in Section 2.3.

Verifying solutions

You can **verify** a solution to a differential equation by substituting the expressions for y and $\dfrac{dy}{dx}$ into the original equation.

Example 2.1

Find the general solution to the differential equation $x\dfrac{dy}{dx} + 5y = \dfrac{2}{x}$ and verify that this satisfies the differential equation.

> When using CAS for derivatives and substitutions, you should clearly show what you have done. Just writing '$x\dfrac{dy}{dx} + 5y = \dfrac{2}{x}$' would not be sufficient.

Solution

The general solution to $x\dfrac{dy}{dx} + 5y = \dfrac{2}{x}$ is $y = \dfrac{2c + x^4}{2x^5}$.

This can be verified by differentiating.

$$y = \frac{2c + x^4}{2x^5} \implies \frac{dy}{dx} = \frac{-10c - x^4}{2x^6}.$$

> Your CAS should be able to find this solution directly. It might give this as
> $$y = \frac{1}{2x} + \frac{c}{x^5}.$$

Substitute into the differential equation.

$$\text{LHS} = x\frac{dy}{dx} + 5y$$

$$= x\left(\frac{-10 - x^4}{2x^6}\right) + 5\left(\frac{2c + x^4}{2x^5}\right)$$

$$= \frac{2}{x}$$

$$= \text{RHS}$$

Families of particular solutions

The general solution to a first order differential equation will have a constant of integration that appears as a parameter. You can use the techniques from Chapter 1 of this book to plot the family of particular solutions and observe any common features they have.

For example, the differential equation $x\dfrac{dy}{dx} + 5y = \dfrac{2}{x}$ has a general solution $y = \dfrac{2c + x^4}{2x^5}$.

The graphs of the particular solutions for $c = -1$ and $c = 2$ are shown in Figures 2.5 and 2.6.

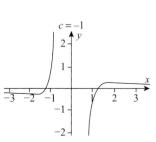

Figure 2.5

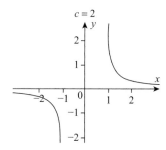

Figure 2.6

By varying c you can observe that for all the curves:

- there is a horizontal asymptote at $y = 0$ and a vertical asymptote at $x = 0$
- they have rotational symmetry about the origin
- there are no points of intersection with the y-axis.

For the cases where $c < 0$:

- there is a point of intersection with both the negative and positive branches of the x-axis
- there is a minimum in the bottom-left quadrant and a maximum in the top-right quadrant.

For the cases where $c \geq 0$:

- there are no points of intersection with either the negative branch or the positive branch of the x-axis
- there are no stationary points.

ACTIVITY 2.2

Find some further examples of first order differential equations.

- Use your CAS to find general and particular solutions for each equation.
- Verify the solutions.
- Vary the values of any parameters and make conjectures about the shapes of the curves of the particular solutions.

Exercise 2.2

① Consider the differential equation $\dfrac{dy}{dx} = 2y - x$.

   (i) Find the general solution.

   (ii) Verify that this satisfies the differential equation.

   (iii) Each particular solution has one of three distinct shapes. Give a value of the constant of integration that generates each of these shapes. Sketch an example of a particular solution in each case and state any common or distinct features of these cases.

② Consider the differential equation $\dfrac{dy}{dx} = xy^2$.

   (i) Find the general solution.

   (ii) Verify that this satisfies the differential equation.

   (iii) Each particular solution has one of three distinct shapes. Give a value of the constant of integration that generates each of these shapes. Sketch an example of a particular solution in each case and state any common or distinct features of these cases.

3 Numerical solutions of differential equations

There are many first order differential equations that cannot be solved analytically. For example,

$$\frac{dy}{dx} = y^2 - \frac{x}{2}.$$

The tangent field for this differential equation is shown in Figure 2.7.

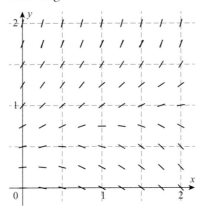

Figure 2.7

In this section, you will meet some numerical methods for solving differential equations such as this. These methods will give you approximations to points on solution curves but will not give you the equations for these curves.

The Euler method

The **Euler method** for solving a differential equation numerically is based on approximating the solution curve (which you cannot find exactly) by a sequence of straight lines. For the differential equation $\frac{dy}{dx} = f(x, y)$, with initial conditions $x = x_0$, $y = y_0$, you can find an approximation of the value of y when $x = x_n$. This is achieved, for a step length of h, by performing n applications of the iteration

The derivation of the method is given on page 29.

$$y_{n+1} = y_n + h f(x_n, y_n)$$

Examples of spreadsheets using the Euler method to approximate the value at $x = 1.5$ of the particular solution of the differential equation $\frac{dy}{dx} = y^2 - \frac{x}{2}$ that passes through the point (1,1) are shown in Figures 2.8 and 2.9. Approximations using step sizes of $h = 0.1$ and $h = 0.05$ are given.

$h = 0.1$

Values of 0.1, 1 and 1 have been entered into cells A2, B2 and C2.

The formula =B2+A$2 has been used in B3.

The formula =C2+A$2*D2 has been used in C3.

The formula =C2^2-B2/2 has been used in D2.

The formulae for the rest of the cells have been obtained by 'filling down'.

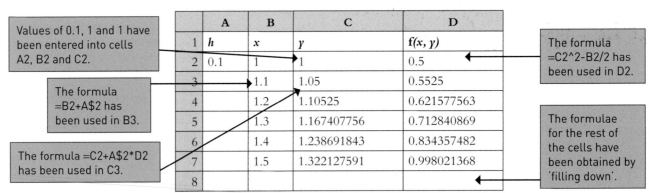

	A	B	C	D
1	*h*	*x*	*y*	f(*x, y*)
2	0.1	1	1	0.5
3		1.1	1.05	0.5525
4		1.2	1.10525	0.621577563
5		1.3	1.167407756	0.712840869
6		1.4	1.238691843	0.834357482
7		1.5	1.322127591	0.998021368
8				

Figure 2.8

Using a step size of $h = 0.1$ gives an approximation of $y = 1.3221$ when $x = 1.5$ (correct to 4 decimal places).

$h = 0.05$

	A	B	C	D
1	*h*	*x*	*Y*	**f(x, y)**
2	0.05	1	1	0.5
3		1.05	1.025	0.525625
4		1.1	1.05128125	0.555192267
5		1.15	1.079040863	0.589329185
6		1.2	1.108507323	0.628788484
7		1.25	1.139946747	0.674478585
8		1.3	1.173670676	0.727502856
9		1.35	1.210045819	0.789210884
10		1.4	1.249506363	0.861266151
11		1.45	1.292569671	0.945736353
12		1.5	1.339856488	1.045215409
13				

Storing the step size *h* in a cell makes it easy to change.

Figure 2.9

Using a step size of $h = 0.05$ gives an approximation of $y = 1.3399$ when $x = 1.5$ (correct to 4 decimal places).

As the approximations for $h = 0.1$ and $h = 0.05$ agree to 1 decimal place, this suggests that $y = 1.3$ when $x = 1.5$ on the true solution curve. However, as the approximations are increasing as h is made smaller, it is possible that the second decimal place could be ≥ 5, therefore you cannot be completely confident in this approximation. In practice, you would need to take further approximations with smaller values of h to be able to decide.

Derivation of the Euler method

In each iteration of the Euler method a straight line is used as an approximation to the solution curve. The gradient of each straight line is taken to be the same as the gradient of the direction indicator in the tangent field at the start of the line.

As an example, take the differential equation

$$\frac{dy}{dx} = y^2 - \frac{x}{2}$$

with initial condition $y = 1$ when $x = 1$. Since you cannot solve the differential equation analytically, you cannot find an equation for the solution curve through the point $(1, 1)$. The Euler method allows you to find an approximation to the actual curve, using the initial conditions and the information you have about the gradient of the curve.

The slope of the direction indicator at the point $(1, 1)$ is given by

$$\frac{dy}{dx} = 1^2 - \frac{1}{2}$$
$$= 0.5$$

Figure 2.10 shows a line segment AB starting at point A (1, 1) with gradient 0.5. The broken line represents the actual (but unknown) solution curve through (1, 1). You can see that the line segment is a tangent to this solution curve at (1, 1). This line segment is the first part of your approximate curve. To find the next part of the curve you need a new starting point.

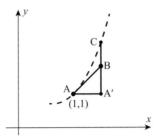

Figure 2.10

After a step size h in the x direction, the exact solution is an unknown point C, and the approximation using the straight line segment is at B. The error in using B instead of C is BC.

If $h = 0.1$, then you can see that in triangle AA'B:

A'B = AA' × gradient of tangent AB

= 0.05.

The approximations for the coordinates of point B are $x = 1.1$ and $y = 1.05$. The next line segment starts from B (1.1, 1.05) and has the same gradient as the direction indicator at B. It is shown as the line BD in Figure 2.11.

The gradient of BD is $1.05^2 - \dfrac{1.1}{2} = 0.5525$.

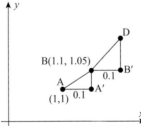

Figure 2.11

The coordinates of point D can be obtained from triangle BB'D using the fact that the gradient of BD is 0.5525.

B'D = 0.1 × 0.5525

= 0.05525

D is therefore the point (1.2, 1.10525).

This method can be repeated so that given a differential equation $\dfrac{dy}{dx} = f(x, y)$, with initial conditions $x = x_0$, $y = y_0$, you can find an approximation of the value of y when $x = x_n$. This is achieved, for a step length of h, by performing n applications of the iteration:

$$y_{n+1} = y_n + h\,f(x_n, y_n)$$

You will realise that drawing a sequence of straight lines from a tangent field is a rather painstaking and slow method. The advantage of the Euler method is that you can formulate a simple rule to generate the coordinates of successive points numerically as a table of values. This means that the process can easily be programmed into a spreadsheet.

The Euler method is an example of a **step-by-step method** because you proceed one small step at a time to find a solution. For each new step, you use the results of the last step. Figure 2.12 shows an example of the Euler method over several steps. Usually, the further you move away from the starting point the larger the error between the sequence of line segments and the actual solution curve. The error can usually be reduced by taking smaller steps.

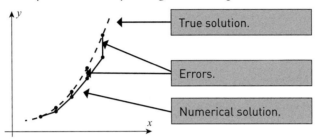

True solution.

Errors.

Numerical solution.

Figure 2.12

ACTIVITY 2.3

Construct spreadsheets like those in the previous examples that use the Euler method to approximate the value at $x = 1.5$ of the particular solution of the differential equation $\dfrac{\mathrm{d}y}{\mathrm{d}x} = y^2 - \dfrac{x}{2}$ that passes through the point $(1, 1)$.

How small does the step size h need to be for you to be confident that the approximation is accurate to 2 decimal places?

Modifying the Euler method

When solving differential equations numerically it is desirable to have methods that converge on accurate solutions as quickly as possible. To do this, modifications to existing methods are often sought that will improve efficiency.

For example, the Euler method can be modified by using the following formulae.

$$k_1 = hf(x_n, y_n)$$
$$k_2 = hf(x_n + h, y_n + k_1)$$
$$y_{n+1} = y_n + \frac{1}{2}(k_1 + k_2)$$

The derivation of the method is given on page 32.

An example of a spreadsheet using this modification of the Euler method to approximate the value at $x = 1.5$ of the particular solution of the differential equation $\dfrac{\mathrm{d}y}{\mathrm{d}x} = y^2 - \dfrac{x}{2}$ that passes through the point $(1, 1)$ is shown in Figure 2.13. A step size of $h = 0.1$ has been used.

Values of 0.1, 1 and 1 have been entered into cells A2, B2 and C2.

The formula =B2+A$2 has been used in B3.

The formulae for the rest of the cells have been obtained by 'filling down'.

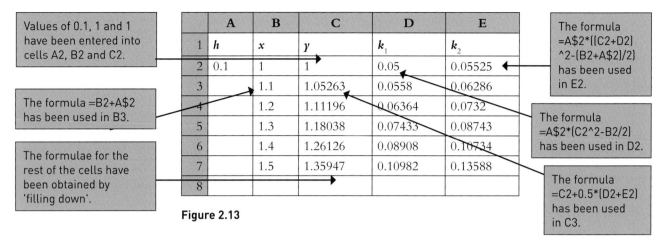

The formula =A$2*((C2+D2)^2-(B2+A$2)/2) has been used in E2.

The formula =A$2*(C2^2-B2/2) has been used in D2.

The formula =C2+0.5*(D2+E2) has been used in C3.

Figure 2.13

Using a step size of $h = 0.1$ gives an approximation of $y = 1.3595$ when $x = 1.5$ (correct to 4 decimal places).

Derivation of this modification of the Euler method

In Figure 2.12, you saw how the numerical solution diverged from the true solution at each step. This error occurs because the gradient of the solution curve over the full length of each interval from (x_n, y_n) to (x_{n+1}, y_{n+1}) is being modelled by the gradient of the direction indicator at the left-hand point of this interval. This modified Euler method attempts to obtain a more accurate approximation of y_{n+1} by using a more representative value for the gradient over this interval.

In Figure 2.14, you can see the tangents to a curve, with $\dfrac{dy}{dx} = f(x, y)$, drawn at two points A and B, representing (x_n, y_n) and (x_{n+1}, y_{n+1}). By taking the mean of the gradients of these tangents we can obtain a more representative approximation for the gradient of the chord over the interval from (x_n, y_n) to (x_{n+1}, y_{n+1}).

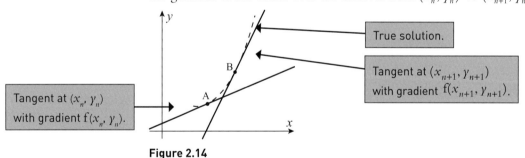

True solution.

Tangent at (x_{n+1}, y_{n+1}) with gradient $f(x_{n+1}, y_{n+1})$.

Tangent at (x_n, y_n) with gradient $f(x_n, y_n)$.

Figure 2.14

The mean of the gradients gives an approximation of the gradient over the interval from (x_n, y_n) to (x_{n+1}, y_{n+1}) of

$$\frac{1}{2}\left(f(x_n, y_n) + f(x_{n+1}, y_{n+1})\right)$$

However, you do not know the value y_{n+1} to substitute into this. You can use the Euler method with a step size of h to approximate the value of y_{n+1} to be $y_n + h f(x_n, y_n)$. Substituting this into the expression for the approximation of the gradient gives

$$\frac{1}{2}\left(f(x_n, y_n) + f(x_{n+1}, y_n + h f(x_n, y_n))\right).$$

This expression can be used to approximate the next value of y for a step size of h.

$$y_{n+1} = y_n + h \times \frac{1}{2}\left(\text{f}\left(x_n, y_n\right) + \text{f}\left(x_{n+1}, y_n + h\,\text{f}\left(x_n, y_n\right)\right)\right)$$

For ease of programming, this is often split into separate expressions with $k_1 = h\text{f}(x_n, y_n)$ and $k_2 = h\text{f}(x_n + h, y_n + k_1)$.

ACTIVITY 2.4

Construct spreadsheets as per the previous example that use this modification of the Euler method to approximate the value at $x = 1.5$ of the particular solution of the differential equation $\dfrac{dy}{dx} = y^2 - \dfrac{x}{2}$ that passes through the point $(1, 1)$.

How small does the step size h need to be for you to be confident that the approximation is accurate to 3 decimal places?

The Runge–Kutta method of order 4

The modification to the Euler method discussed previously is part of a family of methods for solving differential equations numerically that are known as Runge–Kutta methods. The most commonly used of these is the Runge–Kutta method of order 4, and this is often just referred to as the Runge–Kutta method. In practice, this method gives a good balance between accuracy and convenience of programming.

The **Runge–Kutta method of order 4** for solving the differential equation with $\dfrac{dy}{dx} = \text{f}(x, y)$, uses the formulae:

$$k_1 = h\text{f}(x_n, y_n)$$

$$k_2 = h\text{f}\left(x_n + \frac{h}{2}, y_n + \frac{k_1}{2}\right)$$

The background to the method is given on page 34.

$$k_3 = h\text{f}\left(x_n + \frac{h}{2}, y_n + \frac{k_2}{2}\right)$$

$$k_4 = h\text{f}(x_n + h, y_n + k_3)$$

$$y_{n+1} = y_n + \frac{1}{6}(k_1 + 2k_2 + 2k_3 + k_2).$$

An example of a spreadsheet using the Runge–Kutta method of order 4 to approximate the value at $x = 1.5$ of the particular solution of the differential equation $\dfrac{dy}{dx} = y^2 - \dfrac{x}{2}$ that passes through the point $(1, 1)$ is shown in Figure 2.15. A step size of $h = 0.1$ has been used.

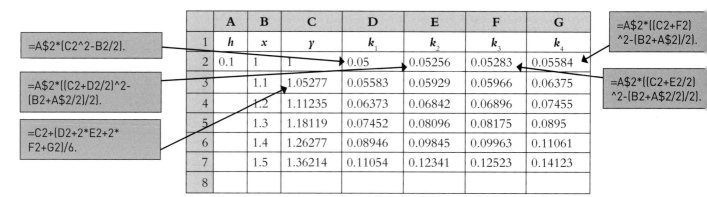

	A	B	C	D	E	F	G
1	h	x	y	k₁	k₂	k₃	k₄
2	0.1	1	1	0.05	0.05256	0.05283	0.05584
3		1.1	1.05277	0.05583	0.05929	0.05966	0.06375
4		1.2	1.11235	0.06373	0.06842	0.06896	0.07455
5		1.3	1.18119	0.07452	0.08096	0.08175	0.0895
6		1.4	1.26277	0.08946	0.09845	0.09963	0.11061
7		1.5	1.36214	0.11054	0.12341	0.12523	0.14123
8							

The formula boxes read:

=A$2*(C2^2-B2/2).

=A$2*((C2+D2/2)^2-(B2+A$2/2)/2).

=C2+(D2+2*E2+2*F2+G2)/6.

=A$2*((C2+F2)^2-(B2+A$2)/2).

=A$2*((C2+E2/2)^2-(B2+A$2/2)/2).

Figure 2.15

Using a step size of $h = 0.1$ gives an approximation of $y = 1.3621$ when $x = 1.5$ (correct to 4 decimal places).

The Runge–Kutta method can be thought of as a weighted average of different approximations for the change in the y value over an interval.

Background to the Runge–Kutta method of order 4

In each iteration of the Runge–Kutta method of order 4, a quadratic curve is used as an approximation to the solution curve. Figure 2.16 shows a quadratic curve being used to approximate a true solution curve. Three points are required to define a quadratic curve, so over an interval with width h the points at $x = x_n$, $x = x_n + \dfrac{h}{2}$ and $x = x_n + h$ can be used, shown by points A, B and C.

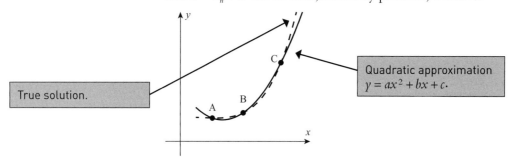

Quadratic approximation $y = ax^2 + bx + c.$

True solution.

Figure 2.16

The Runge–Kutta method of order 4 is based on Simpson's rule for using a quadratic curve to approximate an integral over an interval of width h using the y values of the curve at A, B and C: the points where $x_A = x_n$, $x_B = x_n + \dfrac{h}{2}$ and $x_C = x_n + h$.

$$\int_{x_A}^{x_C} y \, dy = \frac{h}{6}(y_A + 4y_B + y_C)$$

Simpson's rule is covered in full in *MEI Numerical Methods.*

Differentiating this gives an expression for the change in y from A to C based on the derivatives at A, B and C. The derivatives at B and C are not known but can be approximated based on successive applications of the Euler method.

ACTIVITY 2.5

Construct a spreadsheet like in the previous example that uses the Runge–Kutta method of order 4 to approximate the value at $x = 1.5$ of the particular solution of the differential equation $\dfrac{dy}{dx} = y^2 - \dfrac{x}{2}$ that passes through the point $(1,1)$.

How small does the step size h need to be for you to be confident that the approximation is accurate to 5 decimal places?

Variations of these methods

This section provides a brief introduction to the topic of numerical solutions of differential equations. In practice, when solving differential equations numerically, methods are sought that are both efficient and accurate. A large number of variations of the methods described before are available and are used in different situations depending on the type of differential equation to be solved and the technology available.

For example, one variation on the modification to the Euler method discussed previously is to use the following formulae for k_1, k_2 and y_{n+1}.

$$k_1 = hf(x_n, y_n)$$

$$k_2 = hf\left(x_n + \frac{2h}{3}, y_n + \frac{2k_1}{3}\right)$$

$$y_{n+1} = y_n + \frac{1}{4}(k_1 + 3k_2)$$

You should be able to construct spreadsheets for variations of any of the standard methods, given the formulae.

Exercise 2.3

① The value of y when $x = 2$ is required on the solution to the differential equation $\dfrac{dy}{dx} = \sqrt{y} - \dfrac{x}{\sqrt{y}}$ that passes through the point $(1, 3)$.

(i) Construct a spreadsheet to approximate the value of y for the modification of the Euler method with step size h, as follows.

$$k_1 = hf(x_n, y_n)$$

$$k_2 = hf\left(x_n + \frac{2h}{3}, y_n + \frac{2k_1}{3}\right)$$

$$y_{n+1} = y_n + \frac{1}{4}(k_1 + 3k_2)$$

State the formulae you have used in your spreadsheet.

(ii) Use your spreadsheet to approximate the value of y to 5 decimal places when $x = 2$ using a step size of:

(a) $h = 0.1$

(b) $h = 0.05$

(c) $h = 0.025$.

(iii) Give the value of y to an appropriate degree of accuracy and explain your reasoning.

② The value of y when $x = 1$ is required on the solution to the differential equation $\dfrac{dy}{dx} = \sin x + \cos y$ that passes through the point $(0, 1)$.

  (i) Construct a spreadsheet to approximate the value of y for the Runge–Kutta method of order 4 with step size h, as follows.

$$k_1 = hf(x_n, y_n)$$

$$k_2 = hf\left(x_n + \frac{h}{2}, y_n + \frac{k_1}{2}\right)$$

$$k_3 = hf\left(x_n + \frac{h}{2}, y_n + \frac{k_2}{2}\right)$$

$$k_4 = hf(x_n + h, y_n + k_3)$$

$$y_{n+1} = y_n + \frac{1}{6}(k_1 + k_2 + k_3 + k_4)$$

State the formulae you have used in your spreadsheet.

  (i) Use your spreadsheet to approximate the value of y to 8 decimal places when $x = 1$ using a step size of:

    (a) $h = 0.1$

    (b) $h = 0.05$

    (c) $h = 0.025$.

  (ii) Give the value of y to an appropriate degree of accuracy and explain your reasoning.

LEARNING OUTCOMES

Now you have finished this chapter, you should be able to:

➤ use software to produce a tangent field for a differential equation

➤ sketch a tangent field for a first order differential equation and be able to interpret it

➤ use software to find analytical solutions to first order differential equations when this is possible

➤ verify a given solution of a differential equation

➤ work with particular solutions and families of particular solutions

➤ construct, adapt or interpret a spreadsheet to solve first order differential equations numerically

➤ solve a given first order differential equation with initial conditions to any required degree of accuracy by repeated application of the Euler method or a modified Euler method

➤ understand that a smaller step length usually gives a more accurate answer

➤ understand that a modified Euler method usually gives a more accurate solution than an Euler method for a given step length

➤ understand the concepts underlying Runge–Kutta methods

➤ solve first-order differential equations using Runge–Kutta methods.

KEY POINTS

1 A tangent field represents a differential equation as a series of short line segments, with the appropriate gradient, over a set grid of points.

2 The analytical solution to a differential equation can be found directly in some cases using CAS.

3 A differential equation can be verified by substituting the expressions for y and $\frac{dy}{dx}$ into the original equation.

4 The general solution of a differential equation can be represented as a family of curves of the particular solutions.

5 The Euler method for approximating the solution to a differential equation $\frac{dy}{dx} = f(x, y)$ uses n iterations of $y_{n+1} = y_n + hf(x_n, y_n)$ for a given step length h. Using a smaller step length usually gives a more accurate approximation.

6 The Euler method can be modified to give a more accurate approximation by using the iteration $y_{n+1} = y_n + \frac{1}{2}(k_1 + k_2)$, where $k_1 = hf(x_n, y_n)$ and $k_2 = hf(x_n + h, y_n + k_1)$.

7 The Runge–Kutta method of order 4 uses the iteration
$y_{n+1} = y_n + \frac{1}{6}(k_1 + 2k_2 + 2k_3 + k_2)$, where $k_1 = hf(x_n, y_n)$, $k_2 = hf\left(x_n + \frac{h}{2}, y_n + \frac{k_1}{2}\right)$, $k_3 = hf\left(x_n + \frac{h}{2}, y_n + \frac{k_2}{2}\right)$ and $k_4 = hf(x_n + h, y_n + k_3)$.

8 The Euler and Runge–Kutta methods can be further modified by changing the weightings of the k terms.

3

> *Mathematics is the queen of sciences and number theory is the queen of mathematics.*
>
> Carl Friedrich Gauss, 1777–1855

Number theory

The topic of number theory is the study of integers and the relationships between them. This chapter is a brief introduction to number theory explored through programming. You will meet some results relating to prime numbers, modular arithmetic and Diophantine equations (equations with integer solutions).

Programming is a particularly useful tool when applied to number theory problems as it can be used to automate the search for integers that have particular properties. In this chapter, you will be expected to write short programs that produce the solutions to problems and then use mathematical skills to analyse these solutions. For the purposes of this book, all programs will be given in Python version 3.6.3.

Unless stated otherwise, all the problems in this chapter will be restricted to integers.

1 Programming

Mathematical operations and functions in Python

Before starting to write programs, it will be useful to explore how this programming language behaves mathematically.

The basic operations $+, -, \times$ and $\div$ can be implemented directly:

$12+3$ returns the value 15.
$12-3$ returns the value 9.
$12*3$ returns the value 36.
$12/3$ returns the value 4.0.

> Note that the value is not formatted as an integer.

To raise a number to a power, a^n, use **a**n** or **pow(a,n)**:

$5**3$ or **pow(5,3)** both return the value 125.

The percentage symbol is used to give the remainder when dividing by a number:

$31\%7$ returns the value 3 as 31 divided by 7 is 4 remainder 3.

Some further mathematical functions are built into the 'math' library in Python. This library needs to be imported before the functions can be used: to do this, type **import math**. Amongst the most useful of these functions are **floor**, **factorial** and **sqrt** (the square root).

math.floor(6.28) returns the value 6.

> This is the value of 6.28 when rounded down to the nearest integer.

math.factorial(6) returns the value 720.

math.sqrt(289) returns the value 17.0.

> $289**(1/2)$ would also give the same result. Note that the value is not formatted as an integer.

> $6! = 1 \times 2 \times 3 \times 4 \times 5 \times 6$
> $= 720.$

Writing short programs

Programs are defined in Python using the **def** command. You should give the program a name and define any inputs.

The following program multiplies a number by seven:

```
def seven(n):
    m=7*n
    return(m)
```

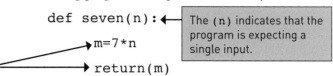
The **(n)** indicates that the program is expecting a single input.

Both of the next two lines are indented by one tab.

seven(5) returns the value **35**.

There are two main techniques that are used within number theory programs:

- systematically looping through all the appropriate values
- checking whether conditions have been satisfied.

To systematically loop through a set of values you can use a **for** command.

The following program prints out the n times table up to the mth term.

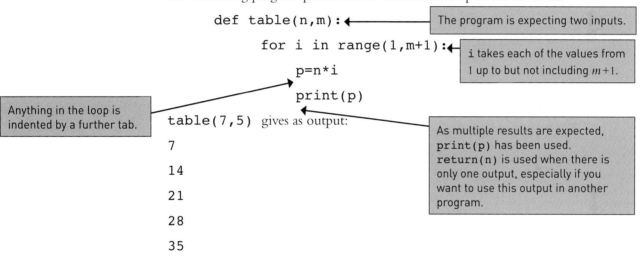

```
def table(n,m):
    for i in range(1,m+1):
        p=n*i
        print(p)
```

The program is expecting two inputs.

i takes each of the values from 1 up to but not including $m+1$.

Anything in the loop is indented by a further tab.

table(7,5) gives as output:

As multiple results are expected, **print(p)** has been used. **return(n)** is used when there is only one output, especially if you want to use this output in another program.

```
7
14
21
28
35
```

To check whether a condition is being met you can use an **if** command.

The following program prints out the 7 times table up to n:

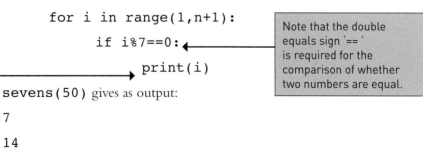

```
def sevens(n):
    for i in range(1,n+1):
        if i%7==0:
            print(i)
```

Note that the double equals sign '==' is required for the comparison of whether two numbers are equal.

Any instruction that is within this indent will be carried out when the condition is true.

sevens(50) gives as output:

```
7
14
21
28
35
42
49
```

In this example, the double equals sign '==' has been used to compare whether two numbers are equal. To check if two numbers are unequal, use '!= '.

An alternative to using `for` and `if` is to use a `while` command. All the examples and problems in this chapter can be solved with `for` and `if`; however, Activity 3.2 on page 42 gives a case where the `while` command is more efficient.

This chapter is not intended to be a detailed introduction to programming in Python. There are many support materials for Python available freely online from sources such as www.python.org.

Exercise 3.1

① (i) Write a program that will find which of the first n positive integers are both 1 more than a multiple of 13 and 2 more than a multiple of 15.

(ii) Use your program to find which of the first 500 positive integers are both 1 more than a multiple of 13 and 2 more than a multiple of 15.

(iii) Adapt your program so that it will find which of the first 500 positive integers are both 1 more than a multiple of 12 and 2 more than a multiple of 15. State the output from this program.

(iv) Show that there are no positive integers that are both 1 more than a multiple of 12 and 2 more than a multiple of 15.

② The numbers 1, 3, 6, 10, 15, 21, 28 and 36 are triangle numbers: numbers that can be written in the form $T_n = \dfrac{n(n+1)}{2}$. 1 and 36 are also square numbers.

(i) Write a program to find which of the first n triangle numbers are also square numbers.

(ii) Use your program to find which of the first 1000 triangle numbers are also square numbers.

(iii) Show that if the nth triangle number T_n is square, then $T_{4n(n+1)}$ is also square. Hence deduce that there are infinitely many numbers that are both triangle and square.

2 Prime numbers

A **prime** number is a positive integer that has exactly two factors: 1 and itself. The first eight prime numbers are 2, 3, 5, 7, 11, 13, 17 and 19. Finding whether a number is prime can be time-consuming, especially if the number is large. For example, try using just a standard scientific calculator to decide whether 10 051 729 or 20 150 407 is prime.

Any integer greater than 1 that is not prime is known as **composite**.

Testing if a number is prime

A simple algorithm to test whether a positive integer is prime is to test whether it can be divided by any of the integers from 2 up to $\sqrt{n}$. If it cannot be divided by any of these, then the number is prime. The following program shows this.

Discussion point

The line
`m=int(n**(1/2))`
finds the largest integer
$\leq \sqrt{n}$.

→ Why is it only
necessary to check up
to the largest integer
less than $\leq \sqrt{n}$?

```
def primetest(n):
    if n==1:
            return(0)
    primetrue=1
    m=int(n**(1/2))
    for i in range(2,m+1):
        if n%i==0:
                primetrue=0
    return(primetrue)
```

As 1 is not a prime number,
but does not have any
other factors, it needs to be
treated as a special case.

ACTIVITY 3.1

Write a more efficient program that will test whether a number is prime.
Compare your program to the previous one for some larger primes (e.g. 15 digits
or more).

Unique prime factorisation

Every integer greater than 1 can be expressed uniquely as the product of powers of
prime numbers. The formal name for this property is **the fundamental theorem
of arithmetic**. A full proof of the result is beyond the scope of this book but is
considered informally in the next discussion point.

Discussion point

→ Why is it not possible
for there to be distinct
primes p_1, p_2, q_1 and q_2
such that $p_1 p_2 = q_1 q_2$?

For example:

$72 = 2^3 \times 3^2$, $1746 = 2 \times 3^2 \times 97$, $1009 = 1009$ (as 1009 is prime).

This result is often presented as the fact that n can be expressed uniquely as

$n = p_1^{a_1} p_2^{a_2} p_3^{a_3} \ldots p_m^{a_m}$, where $p_1, p_2, \ldots p_m$ are distinct primes and $a_1, a_2, \ldots a_m$ are
positive integers.

Greatest common divisor

The **greatest common divisor (GCD)** of two positive integers a and b is the
largest integer c such that c divides a and c divides b.

For example, the greatest common divisor of 24 and 60 is 12, and the greatest
common divisor of 72 and 175 is 1.

The greatest common divisor is sometimes known as the **highest common
factor (HCF)**.

Example 3.1

(i)   Write a program that will find the greatest common divisor of two positive
integers m and n, where $m \leq n$.

(ii) Use your program to find the greatest common divisors of:

(a) 36 and 100
(b) 25 and 169
(c) 7056 and 17 424.

(iii) Show that if m and n are square numbers, then the greatest common divisor of
m and n is a square number.

return(g) is used
because the output from
gcd(m,n) is a single
number that can be called
from other programs.
print(g) would work in
this example, but you will
see a use of this program
on page 47 that requires
return(g).

Solution

(i)
```
def gcd(m,n):
    for i in range (1,m+1):
        if n%i==0 and m%i==0:
            g=i
    return(g)
```

(ii)

   (a) The greatest common divisor of 36 and 100 is 4.

   (b) The greatest common divisor of 25 and 169 is 1.

   (c) The greatest common divisor of 7056 and 17 424 is 144.

(iii) If m and n are square numbers, then $m = p_1^{a_1} p_2^{a_2} p_3^{a_3} \ldots$ and

$n = p_1^{b_1} p_2^{b_2} p_3^{b_3} \ldots$, where all values of $a_1, a_2, a_3 \ldots$ and $b_1, b_2, b_3 \ldots$ are zero or positive even numbers. For each prime p_i, the highest power of p_i that is both a factor of m and n is an even number. Therefore the greatest common divisor of m and n is the product of even powers of primes and hence a square number.

ACTIVITY 3.2

One algorithm for finding the greatest common divisor of two integers is the Euclidean algorithm. A program for this algorithm is provided here. Compare the efficiency of this program with that of the program given in Example 3.1 for some large values of a and b.

```
def gcd(a,b):
    m=a
    n=b
    while n!=0:
        k=int(m/n)
        oldn=n
        n=m-k*n
        m=oldn
    return(m)
```

Co-prime integers

Positive integers a and b are **co-prime** if the greatest common divisor of a and b is 1.

If two integers a and b are not co-prime then they can be written as $a = kc$ and $a = kd$ where c, d and k are positive integers and $k > 1$.

Example 3.2

(i) Write a program that will find all the possible pairs of positive integers x and y that are solutions to the equation $ax + by = n$, where a, b and n are also positive integers.

(ii) Use your program to find all the possible pairs of positive integers x and y that are solutions to:

(a) $4x + 10y = 46$

(b) $91x + 97y = 3912$

(c) $91x + 208y = 3913$

(d) $91x + 208y = 3914$.

(iii) Show that if a and b are not co-prime and $ax + by = n$ has solutions, then n is not co-prime with either a or b.

Solution

(i)
```
def solve(a,b,n):
    p=int(n/a)
    q=int(n/b)
    for x in range(1,p+1):
        for y in range(1,q+1):
            if a*x+b*y==n:
                print(x,y)
```

These are the maximum possible values of x and y.

This program does an exhaustive search of all possible combinations. A more efficient method based on checking (n-a*x)%y==0 could be used.

print(x,y) is required as it is possible to have multiple solutions.

(ii) The possible pairs of solutions are:

(a) $4x + 10y = 46$ has solutions $x = 4$, $y = 3$ and $x = 9$, $y = 1$

(b) $91x + 97y = 3912$ has solutions $x = 27$, $y = 15$

(c) $91x + 208y = 3913$ has solutions $x = 11$, $y = 14$ and $x = 27$, $y = 7$

(d) $91x + 208y = 3914$ has no solutions.

Examples (a) and (c) are the only ones that meet the conditions of (iii).

(iii) If a and b are not co-prime, then they can be written as $a = kc$ and $a = kd$, where $k > 1$.

$ax + by = n$ can be written as $kcx + kdy = n$ or $k(cx + dy) = n$.
Therefore, any value of n for which there are solutions must be a multiple of k. Hence a and n share a factor of k and are not co-prime, and likewise b and n share a factor of k and are not co-prime.

① (i) Write a program that will find all of the cases where $2^k - 1$ is prime for $1 < k \leq n$.

(ii) Use your program to find all of the cases where $2^k - 1$ is prime for $1 < k \leq 15$, giving the values of k and $2^k - 1$.

(iii) A perfect number is a positive integer that is equal to the sum of its divisors other than itself: e.g. 6 is perfect because $1 + 2 + 3 = 6$. Show that if $2^k - 1$ is prime, then $2^{k-1}(2^k - 1)$ is perfect.

(iv) Hence give the value of any perfect numbers predicted by your answer to (ii).

② (i) Explain why, for two positive integers m and n, their product mn is an upper bound for the lowest common multiple of m and n.

(ii) Write a program to find the lowest common multiple of two numbers m and n.

(iii) Use your program to find the lowest common multiple of $1\,600\,306$ and $1\,490\,696$.

(iv) Show that if m and n are not co-prime, then the lowest common multiple of m and n is strictly less than mn.

3 Congruences and modular arithmetic

Modular arithmetic

Modular arithmetic is a branch of arithmetic where you use a finite number of integers and, when counting, you return to the start of the list after you have reached the end of it. You can think of the numbers as looping round.

One common example of modular arithmetic is in telling the time. For example, the time that is 5 hours after 10 o'clock is 3 o'clock (using the 12-hour clock): that is, you are only interested in the remainder when it is divided by 12. Using the conventions of modular arithmetic you would say '15 is congruent to 3 modulo 12'. This is written as

$$15 \equiv 3 \pmod{12}$$

An 'equation' written in modular arithmetic is known as a **congruence**. The congruence $5n + 7 \equiv 4 \pmod{11}$ has the solution $n = 6$ because $37 \equiv 4 \pmod{11}$. Note that $n = 17$ and $n = -5$ are also valid, but the result is usually written as $n = 6$ as the convention when working mod m is to give results as an integer n, where $0 \leq n < m$.

It is possible to use methods of programming to search for the solutions to congruences, as can be seen in the following example.

Example 3.3

(i) Write a program that will solve the congruence $an + b \equiv c \pmod{p}$.

(ii) Use your program to solve the congruences:

    (a) $12n + 60 \equiv 1 \pmod{97}$

    (b) $49n + 5 \equiv 90 \pmod{97}$

    (c) $n + 10 \equiv 83 \pmod{97}$.

(iii) Show that the congruences $49n + 5 \equiv 90 \pmod{97}$ and $n + 10 \equiv 83 \pmod{97}$ are equivalent.

Solution

(i)
```
def congruence(a,b,c,p):
    for n in range(0,p):
        if (a*n+b)%p==c:
            print(n)
```

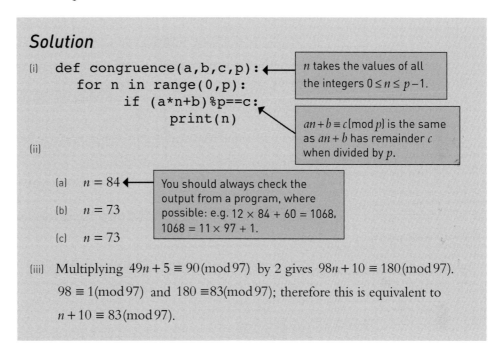

n takes the values of all the integers $0 \leq n \leq p - 1$.

$an + b \equiv c \pmod{p}$ is the same as $an + b$ has remainder c when divided by p.

(ii)

    (a) $n = 84$

    (b) $n = 73$

    (c) $n = 73$

You should always check the output from a program, where possible: e.g. $12 \times 84 + 60 = 1068$, $1068 = 11 \times 97 + 1$.

(iii) Multiplying $49n + 5 \equiv 90 \pmod{97}$ by 2 gives $98n + 10 \equiv 180 \pmod{97}$.

$98 \equiv 1 \pmod{97}$ and $180 \equiv 83 \pmod{97}$; therefore this is equivalent to $n + 10 \equiv 83 \pmod{97}$.

When working in modulo n, for small values of n, it can often be useful to consider all possible cases modulo n.

For example, it is possible to use proof by exhaustion to prove that for all positive integers n, either $n^2 \equiv 0 \pmod{4}$ or $n^2 \equiv 1 \pmod{4}$.

There are only four possible cases when working modulo 4:

• $n \equiv 0 \pmod{4}$: $0^2 = 0$ and therefore $n^2 \equiv 0 \pmod{4}$

• $n \equiv 1 \pmod{4}$: $1^2 = 1$ and therefore $n^2 \equiv 1 \pmod{4}$

• $n \equiv 2 \pmod{4}$: $2^2 = 4$ and therefore $n^2 \equiv 0 \pmod{4}$

• $n \equiv 3 \pmod{4}$: $3^2 = 9$ and therefore $n^2 \equiv 1 \pmod{4}$.

ACTIVITY 3.3

Show this result by expanding the expressions $(4k)^2$, $(4k + 1)^2$, $(4k + 2)^2$ and $(4k + 3)^2$.

Fermat's little theorem

By considering the successive powers of some integers modulo 5, it can be observed that:

$2^3 = 8 \equiv 3(\mod 5)$.

$2^2 \equiv 4, 2^3 \equiv 3, 2^4 \equiv 1, 2^5 \equiv 2$

$3^2 \equiv 4, 3^3 \equiv 2, 3^4 \equiv 1, 3^5 \equiv 3$

$4^2 \equiv 1, 4^3 \equiv 4, 4^4 \equiv 1, 4^5 \equiv 4$.

This suggests that $a^5 \equiv a(\mod 5)$, and this result generalises to Fermat's little theorem:

$a^p \equiv a(\mod p)$, where p is prime.

Fermat's little theorem can be proved by induction, as shown.

For the case $a = 0$:

$0^p \equiv 0(\mod p)$, so it is true for $a = 0$.

Discussion point

→ How can you show that if p is prime, then p is a factor of $_pC_n$ for $1 \leq n < p$?

Assume that it is true for $a = k$, so $k^p \equiv k(\mod p)$.

For $a = k + 1$:

$(k+1)^p = k^p + {_pC_1}k^{p-1} + {_pC_2}k^{p-2} + \ldots + {_pC_{p-1}}k + 1$

If p is prime, then p is a factor of $_pC_n$ for $1 \leq n < p$.

$\equiv k + 0 + 0 + \ldots + 0 + 1(\mod p)$

$\equiv k + 1(\mod p)$

Thus, if the result is true for $a = k$, then it is true for $a = k + 1$. Since it is true for $a = 0$, then, by induction, it is also true for all positive integers.

Example 3.4

(i) Write a program that will test whether a positive integer n satisfies $a^n \equiv a(\mod n)$ for all values of a, where $0 \leq a < n$.

(ii) Use your program to test whether $a^n \equiv a(\mod n)$ for $0 \leq a < n$ for all values of a, where $0 \leq a < n$, when:

(a) $n = 7$

(b) $n = 91$

(c) $n = 191$

(d) $n = 561$.

Solution

(i)
```
def flt(n):
    flttrue=1
    for i in range(0,n):
        if i**n%n!=i:
            flttrue=0
    return(flttrue)
```

(ii) Running the program gives the results:

(a) `flt(7)=1`, $a^n \equiv a(\bmod n)$ for $0 \leq a < n$ is true for $n = 7$

(b) `flt(91)=0`, $a^n \equiv a(\bmod n)$ for $0 \leq a < n$ is false for $n = 91$

(c) `flt(191)=1`, $a^n \equiv a(\bmod n)$ for $0 \leq a < n$ is true for $n = 191$

(d) `flt(561)=1`, $a^n \equiv a(\bmod n)$ for $0 \leq a < n$ is true for $n = 561$.

Discussion point

→ How can you derive the second version of Fermat's little theorem from the first version?

If p is not a factor of a, then Fermat's little theorem can be written

$a^{p-1} \equiv 1(\bmod p)$, where p is prime.

EXTENSION ACTIVITY

Composite numbers n that satisfy $a^n \equiv a(\bmod n)$ for $0 \leq a < n$, such as 561, are known as Carmichael numbers. Write a program to find all the Carmichael numbers n where $n < 10000$. Can you find any common features of these numbers?

Euler's totient function and Euler's theorem

On page 42 you met the idea of pairs of numbers being co-prime. For a given value of n, some of the numbers less than n are co-prime with n, and it can be useful to know how many of these there are.

Euler's totient function $\varphi(n)$ counts the number of positive integers less than n that are co-prime with n.

For example:

$\varphi(7) = 6$ because $1, 2, 3, 4, 5$ and 6 are all co-prime with 7

$\varphi(12) = 4$ because $1, 5, 7$ and 11 are all co-prime with 12.

If a number n is prime, then $\varphi(n) = n - 1$ as all of the integers less than n are co-prime with n.

If a number n is composite, then $\varphi(n) < n - 1$ as some of the integers less than n have a common factor with n.

You can call the result of one of the programs for the greatest common divisor on page 42, defined as `gcd(m,n)`, to create a program that will find $\varphi(n)$.

```
def etf(n):
    t=0
    for i in range(1,n):
        if gcd(i,n)==1:
            t=t+1
    return(t)
```

One application of Euler's totient function is Euler's theorem. **Euler's theorem** states that if a and n are co-prime positive integers, then

$a^{\varphi(n)} \equiv 1(\text{mod } n)$.

For the cases where n is prime, Euler's theorem is the same as Fermat's little theorem.

For the cases where n is composite, Euler's theorem can be demonstrated by considering all the possible numbers that are co-prime with n.

For example, 5 and 12 are co-prime and $\varphi(12) = 4$.

$5^4 = 625$

$= 12 \times 156 + 1$

and hence $5^{\varphi(12)} \equiv 1(\text{mod } 12)$.

The four values of m ($m < 12$), where m and 12 are co-prime, are $m = 1, 5, 7$ and 11. Considering $5m$ modulo 12 for each value (as $a = 5$):

$5 \times 1 \equiv 5 \ (\text{mod } 12), 5 \times 5 \equiv 1 \ (\text{mod } 12), 5 \times 7 \equiv 11 \ (\text{mod } 12)$ and $5 \times 11 \equiv 7 \ (\text{mod } 12)$.

Equating the product of all four left-hand sides of these congruences with all four right-hand sides gives

$5^4 \times 1 \times 5 \times 7 \times 11 \equiv 1 \times 5 \times 7 \times 11 \ (\text{mod } 12)$.

This can be simplified to
$5^4 \equiv 1 \ (\text{mod } 12)$. ←

> Each of the four values of m has appeared once as the right-hand side of one of these congruences.

Discussion point

→ In Activity 3.4, does each value of m always appear exactly once as a value of $am \ (\text{mod } n)$?

ACTIVITY 3.4

Find some other pairs of values of a and n that are co-prime positive integers, with n being composite. In each case:

- find $\varphi(n)$
- list the $\varphi(n)$ values of m that are co-prime with m, where $m < n$
- find $am \ (\text{mod } n)$ for each such value.
- show that each value of m appears once as a value of $am \ (\text{mod } n)$.
- deduce that $a^{\varphi(n)} \equiv 1(\text{mod } n)$.

Wilson's theorem

The product of all the positive integers up to and including n is given by the factorial function $n! = 1 \times 2 \times 3 \times \ldots \times n$.

> Take care when programming involving the factorial function as ! = is used for ≠.

Some values of $(p - 1)!(\text{mod } p)$ are given:

$4! = 24$	$6! = 720$	$10! = 3628800$	$12! = 479001600$
$\equiv 4 \ (\text{mod } 5)$	$\equiv 6 \ (\text{mod } 7)$	$\equiv 10 \ (\text{mod } 11)$	$\equiv 12 \ (\text{mod } 13)$

These values suggest that if p is prime, then $(p - 1)! \equiv p - 1(\text{mod } p)$.

The following program gives all the values of n where $(n-1)! \equiv n-1 \pmod{n}$:

```
import math
def wilson(n):
    for i in range(2,n+1):
        if math.factorial(i-1)%i==(i-1):
            print(i)
```

Running the code to find the output of `wilson(30)` gives: 2, 3, 5, 7, 11, 13, 17, 19, 23 and 29, which are all prime numbers. This suggests that the result holds in the opposite direction too, so that if $(p-1)! \equiv p-1 \pmod{p}$, then p is prime.

The result is known as **Wilson's theorem** and is usually presented as

$(p-1)! \equiv -1 \pmod{p} \Leftrightarrow p$ is prime

Examining the value of 12! modulo 13 can give an insight into why this might be true in general for all primes:

$12! = 1 \times 2 \times 3 \times 4 \times 5 \times 6 \times 7 \times 8 \times 9 \times 10 \times 11 \times 12$

$\quad = 1 \times (2 \times 7) \times (3 \times 9) \times (4 \times 10) \times (5 \times 8) \times (6 \times 11) \times 12$

$\quad \equiv 1 \times 1 \times 1 \times 1 \times 1 \times 1 \times 12 \pmod{13}$

$\quad \equiv 12 \pmod{13}$

> Each of these products of a pair is congruent to 1 modulo 13, e.g. $66 \equiv 1 \pmod{13}$.

Discussion point

→ Why is $(n-1)!$ never congruent to $n-1$ modulo n when n is composite?
Hint: consider the case where n is the square of a prime number separately.

ACTIVITY 3.5

For some different values of p, where p is prime and $p > 3$, show that $2 \times 3 \times \ldots \times (p-2)$ can be written as the product of pairs, each of which is congruent to 1 modulo p. Is this true for any prime value of p?

Exercise 3.3

① (i) Write a program that will find, for a given positive integer n, all the integers a and b such that $a^n + b^n \equiv m \pmod{n}$.

(ii) Use your program to list all the integers a and b such that:

(a) $a^7 + b^7 \equiv 5 \pmod{7}$

(b) $a^{13} + b^{13} \equiv 0 \pmod{13}$.

(iii) Show that, if n is prime, then $a^n + b^n \equiv m \pmod{n} \Rightarrow a + b \equiv m \pmod{n}$.

② This question concerns Euler's totient function $\varphi(n)$: the number of positive integers less than n that are co-prime with n.

(i) Write a program that will list all the positive integers that are co-prime with a given positive integer n.

(ii) Use your program to list all the positive integers that are co-prime with:

(a) 8

(b) 25.

(iii) Given that p is prime, show that $\varphi(p^a) = p^{a-1}(p-1)$. Use this result to verify the values of $\varphi(8)$ and $\varphi(25)$.

4 Diophantine equations

Diophantus was a Greek mathematician who lived in the third century AD. He is known for his work on equations, especially those with integer solutions. Consequently, equations where only integer solutions are sought are called **Diophantine equations**. In many cases, Diophantine equations are further restricted to exclusively positive integer solutions.

For example, the equation $5x + 4y = 34$ has infinitely many solutions when x and y are real numbers but just the two solutions, $x = 2, y = 6$ and $x = 6, y = 1$, when x and y are positive integers.

As solving Diophantine equations requires finding integer solutions, they lend themselves well to the methods based on number theory techniques.

Pythagorean triples

One of the most studied Diophantine equations is Pythagoras' theorem, $a^2 + b^2 = c^2$, restricted to the cases where a, b and c are positive integers. These are known as **Pythagorean triples**. Some examples of Pythagorean triples are shown in Figure 3.1.

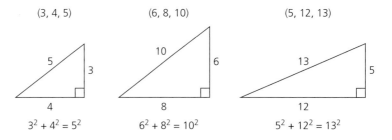

$$(3, 4, 5) \qquad (6, 8, 10) \qquad (5, 12, 13)$$

$$3^2 + 4^2 = 5^2 \qquad 6^2 + 8^2 = 10^2 \qquad 5^2 + 12^2 = 13^2$$

Figure 3.1

Pythagorean triples that do not share a common factor are known as **primitive Pythagorean triples**. For example, $(3, 4, 5)$ and $(5, 12, 13)$ are primitive Pythagorean triples, but $(6, 8, 10)$ is not since $(6, 8, 10)$ can be generated by multiplying the primitive Pythagorean triple $(3, 4, 5)$ by 2. This is explored further in the following example.

Example 3.5

This question concerns Pythagorean triples (a, b, c): solutions to $a^2 + b^2 = c^2$, where a, b and c are positive integers.

(i)   Write a program to find all the Pythagorean triples (a, b, c), where $c < n$.

(ii) Use your program to find all the Pythagorean triples (a, b, c), where $c < 30$.

(iii) State which of these solutions are not primitive Pythagorean triples.

Hint: consider the case where a and b are co-prime but a and c are not.

(iv) Show that, for a Pythagorean triple, if a and b are co-prime, then a and c are co-prime and b and c are co-prime.

Solution

(i)
```
def pythag(n):
    for c in range(1,n):
        for a in range(1,c):
            for b in range(a,c):
                if (a*a+b*b)==(c*c):
                    print(a,b,c)
```

(ii) Solutions are: $(3, 4, 5), (6, 8, 10), (5, 12, 13), (9, 12, 15), (8, 15, 17),$ $(12, 16, 20), (7, 24, 25), (15, 20, 25), (10, 24, 26), (20, 21, 29).$

(iii) The solutions that are not primitive Pythagorean triples are:

$(6,8,10) = 2 \times (3,4,5)$

$(9,12,15) = 3 \times (3,4,5)$

$(12,16,20) = 4 \times (3,4,5)$

$(15,20,25) = 5 \times (3,4,5)$

$(10,24,26) = 2 \times (5,12,13)$

(iv) Assume that a and c are not co-prime. a and c can then be written $a = km$ and $c = kn$, where k is an integer and $k \geq 2$.

$a^2 + b^2 = c^2 \Rightarrow (km)^2 + b^2 = (kn)^2$

$\Rightarrow k^2 n^2 - k^2 m^2 = b^2$

$\Rightarrow k^2(n^2 - m^2) = b^2$

As k^2 is a factor of b^2, then k is a factor of b and hence a and b share a common factor and are not co-prime. This contradicts the given condition that a and b are co-prime, and therefore the assumption that a and c are not co-prime is false: a and c are co-prime. The result follows similarly for b and c.

Pell's equation

Another well-studied Diophantine equation is Pell's equation. **Pell's equation** is the set of Diophantine equations

$$x^2 - ny^2 = 1$$

for which n is a positive integer.

The solution $x = 1$, $y = 0$ is a trivial solution for all values of n; however, the values of x and y are often restricted to positive integers, which excludes this. The relationship between the non-trivial solutions is explored in the following example.

Example 3.6

(i) Write a program to find all the possible positive integer solutions to the equation $x^2 - 2y^2 = 1$, where $x \leq m$ and $y \leq m$.

(ii) Use your program to find all the solutions where $x \leq 100$ and $y \leq 100$.

(iii) By writing the smallest solution to $x^2 - 2y^2 = 1$ in the form $(x + \sqrt{2}y)(x - \sqrt{2}y) = 1$, show how it can be used to generate the other solutions.

Solution

(i)
```
def pell(m):
    for x in range(1,m+1):
        for y in range(1,m+1):
            if x*x-2*y*y==1:
                print(x,y)
```

(ii) $x^2 - 2y^2 = 1$ has the following solutions.

$x = 3, y = 2$

$x = 17, y = 12$ ←

$x = 99, y = 70$

> Always check the output of any program. In this case:
> $9 - 8 = 1$,
> $289 - 288 = 1$,
> $9801 - 9800 = 1$.

(iii) $3^2 - 2 \times 2^2 = 1$ can be written $(3 + \sqrt{2} \times 2)(3 - \sqrt{2} \times 2) = 1$.
Squaring both sides gives $(3 + \sqrt{2} \times 2)^2(3 - \sqrt{2} \times 2)^2 = 1$, which can be simplified to $(17 + \sqrt{2} \times 12)(17 - \sqrt{2} \times 12) = 1$, i.e. $17^2 - 2 \times 12^2 = 1$.
Therefore $x = 17, y = 12$ is also a solution.
Similarly, $(3 + \sqrt{2} \times 2)^3(3 - \sqrt{2} \times 2)^3 = 1$ gives $99^2 - 2 \times 70^2 = 1$, and therefore $x = 99, y = 70$ is a solution.

Discussion point

→ Can all possible solutions to $x^2 - ny^2 = 1$ be generated from the smallest possible solution?

ACTIVITY 3.6

Adapt the code in Example 3.6 so that it will find the solutions to Pell's equation for some other values of n. For each case, show that any solution to $x^2 - ny^2 = 1$ can be used to generate further solutions.

One application of Pell's equation is finding rational approximations to the square roots of positive integers. Rearranging the equation gives $n = \dfrac{x^2 - 1}{y^2}$ and hence $\sqrt{n} \approx \dfrac{x}{y}$, with the approximation being increasingly accurate as x and y increase.

For example, the solutions in the previous example give the following approximations to $\sqrt{2}$ (which is 1.414214 to 6 decimal places):

Discussion point

→ When n is a square number, does Pell's equation $x^2 - ny^2 = 1$ have any solutions for which x and y are positive integers?

$x = 3, y = 2 \Rightarrow \sqrt{2} \approx \dfrac{3}{2}$ or 1.5;

$x = 17, y = 12 \Rightarrow \sqrt{2} \approx \dfrac{17}{12}$ or 1.416667;

$x = 99, y = 70 \Rightarrow \sqrt{2} \approx \dfrac{99}{70}$ or 1.414286.

Other Diophantine equations

The method of creating a program to perform an exhaustive search of integers can be used to solve many Diophantine equations, including variations on Pythagoras' theorem and Pell's equation. This can be seen in the following example.

Example 3.7

(i) Write a program to find all the possible positive integer solutions to the equation $a^2 + b^2 = c^2 - 1$, where $c \le n$.

(ii) Use your program to find all the solutions where $c \le 50$.

(iii) By considering all the possible values of $m^2 \pmod 8$, show that, for any possible solution, either both a and b are congruent to 0 modulo 4 or both a and b are congruent to 2 modulo 4.

Solution

(i)
```
def pythag1(n):
    for c in range(1,n):
        for a in range(1,c):
            for b in range(a,c):
                if (a*a+b*b)==(c*c)-1:
                    print(a,b,c)
```

(ii) The program returns the following triples: $(2, 2, 3)$, $(4, 8, 9)$, $(12, 12, 17)$, $(6, 18, 19)$, $(8, 32, 33)$, $(18, 30, 35)$.

(iii) The values of $m^2 \pmod 8$ are $0^2 \equiv 0$, $1^2 \equiv 1$, $2^2 \equiv 4$, $3^2 \equiv 1$, $4^2 \equiv 0$, $5^2 \equiv 1$, $6^2 \equiv 4$, $7^2 \equiv 1$, i.e. the only possible values are 0, 1 or 4.
Hence the only possible values of $c^2 - 1 \pmod 8$ are 0, 3 or 7.
$a^2 + b^2$ is the sum of two values that are either 0, 1 or 4. The only possible such sums that give a result of 0, 3 or 7 are $0 + 0 \equiv 0$ or $4 + 4 \equiv 0$.
For $0 + 0 \equiv 0$, both a and b are congruent to 0 or 4 modulo 8 and therefore $a \equiv 0 \pmod 4$ and $b \equiv 0 \pmod 4$.
For $4 + 4 \equiv 0$, both a and b are congruent to 2 or 6 modulo 8 and therefore $a \equiv 2 \pmod 4$ and $b \equiv 2 \pmod 4$.

ACTIVITY 3.7

One of the most famous Diophantine equations is the subject of Fermat's last theorem. **Fermat's last theorem** states that there are no solutions to the equation $a^n + b^n = c^n$, where a, b, c and n are positive integers and $n > 2$.
Investigate the 'near misses' to Fermat's last theorem, e.g. solutions to $a^3 + b^3 = c^3 + 1$.

Exercise 3.4

① This question concerns Pythagorean triples (a, b, c): solutions to $a^2 + b^2 = c^2$, where a, b and c are positive integers.

(i) Write a program to find all the Pythagorean triples (a, b, c), where $c < n$ and a is prime.

(ii) Use your program to find all the Pythagorean triples (a, b, c), where $c < 100$ and a is prime.

(iii) By rewriting $a^2 + b^2 = c^2$ as $a^2 = c^2 - b^2$, show that $a = 2$ is not possible in a solution to $a^2 + b^2 = c^2$, where a, b and c are positive integers.

(iv) Hence explain why it is not possible for all three values in a Pythagorean triple to be prime.

② (i) Write a program that will find all the positive integer solutions to the equation $x^2 - ny^2 = -1$, where $x \leq 100$ and $y \leq 100$.

(ii) Use your program to find all the positive integer solutions (where $x \leq m$ and $y \leq m$) to:

(a) $x^2 - 2y^2 = -1$

(b) $x^2 - 4y^2 = -1$

(c) $x^2 - 5y^2 = -1$.

(iii) Show that if $n \equiv 0 \pmod 4$, then $x^2 - ny^2 = -1$ has no solutions.

LEARNING OUTCOMES

Now you have finished this chapter, you should be able to:

➤ write, adapt and interpret programs to solve number theory problems
➤ identify the limitations of a short program and suggest refinements to it
➤ use the unique prime factorisation of natural numbers
➤ solve problems using modular arithmetic
➤ use Fermat's little theorem
➤ use Euler's totient function $\varphi(n)$ and Euler's theorem
➤ use Wilson's theorem
➤ find Pythagorean triples and use related equations
➤ solve Pell's equation and use solutions to solve related problems
➤ solve other Diophantine equations and use solutions to solve related problems.

KEY POINTS

1 To systematically check whether the integers in a given range satisfy given conditions, use a program with a loop and checking statements. This is commonly done with `for` and `if` statements.

2 Every positive integer n, where $n > 1$, can be expressed uniquely as the product of powers of prime numbers: $n = p_1^{a_1} p_2^{a_2} p_3^{a_3} \dots p_m^{a_m}$ where $p_1, p_2, \dots p_m$ are distinct primes and $a_1, a_2, \dots a_m$ are positive integers.

3 The greatest common divisor (GCD) of two positive integers a and b is the largest integer c such that c divides a and c divides b.

4 Positive integers a and b are co-prime if the greatest common divisor of a and b is 1.

5 Modular arithmetic is a branch of arithmetic that uses a finite number of integers and, when counting, returns to the start of the list after reaching the end of it. An 'equation' written in modular arithmetic is known as a congruence.

6 Fermat's little theorem states that $a^p \equiv a \pmod{p}$, where p is prime.

7 Euler's totient function $\varphi(n)$ counts the number of positive integers less than n that are co-prime with n. Euler's theorem states that if a and n are co-prime positive integers, then $a^{\varphi(n)} \equiv 1 \pmod{n}$.

8 Wilson's theorem states that $(p-1)! \equiv -1 \pmod{p} \Leftrightarrow p$ is prime.

9 Equations where only integer solutions are sought are called Diophantine equations.

10 Pythagorean triples are positive integers a, b and c that are solutions to the equation $a^2 + b^2 = c^2$.

11 Pell's equation is the set of Diophantine equations $x^2 + ny^2 = 1$ for which n is a positive integer.

Answers

Chapter 1

Activity 1.1 (Page 2)

You should ensure that you are able to obtain the curves in Figure 1.1 in your software. Select some other curves that you are familiar with and plot these in your software to help you learn how the syntax works and how to move around the axes.

Discussion point (Page 4)

- $y = 3x + c$ is a translation of $y = 3x$ by c units on the positive y-axis.
- $y = x^2 + bx + 4$ is a translation of $y = x^2 + 4$ by

  the vector $\begin{pmatrix} -\frac{b}{2} \\ -\frac{b^2}{4} \end{pmatrix}$.

- $y = e^{kx}$ is a stretch of $y = e^x$ by a factor of $\frac{1}{k}$ in the x-direction.

Function notation is useful for considering transformations. For the graph of $y = f(x)$: translations can be obtained with $y = f(x + a) + b$; stretches can be obtained with $y = c\,f(dx)$; reflections can be obtained with $y = f(-x)$ or $y = -f(x)$. One interesting case to consider is the graph of $y = \ln(kx)$ which can be considered as either a stretch or a translation of the graph of $y = \ln(x)$.

Activity 1.2 (Page 4)

You could try taking a curve whose equation you are familiar with, such as $y = ax^2 + bx + c$, and varying each of the parameters in turn. For each parameter you vary you should be able to describe the resulting transformation and investigate how you can show this mathematically.

Activity 1.3 (Page 5)

You should ensure that you are able to obtain the results on pages 4 and 5 with your software. Use CAS for some other results you are familiar with to help you learn how the syntax works and the format the results will be presented in.

Discussion point (Page 6)

To show the common features:

- $t = 0 \Rightarrow x = 0, y = 0$.

- Substituting in $t = -t_1$.

- Show that $0 < \dfrac{t^2}{1 + t^2} < 1$ for all t.

- Show that $x \to 1$ and $y \to \infty$ as $t \to \infty$.

To show the distinct features:

- Show that case $a = -1$ has two distinct values of t that generate the same point.
- Find the values of t for which $x > 0$ and $y = 0$.
- Show that the curve is defined at $(0, 0)$ and the gradient of the tangent tends to the same limit as the point is approached along either branch and this technique is covered in Section 1.3.

Exercise 1.1 (Page 7)

1 (i)

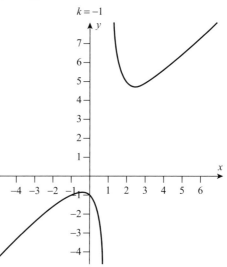

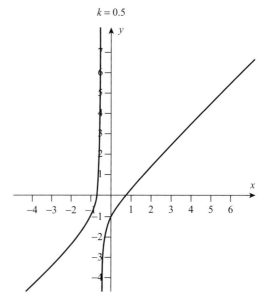

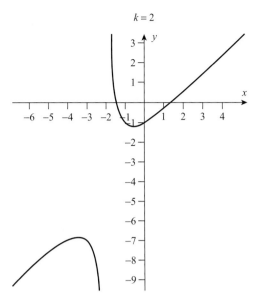

$k = 2$

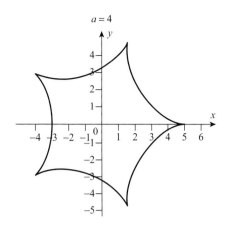

$a = 4$

Common features include: all the curves pass through the point $(a + 1, 0)$; all the curves are symmetrical about the x-axis; all the curves have rotational symmetry order $a + 1$; all the curves are bounded by a circle centred at the origin with radius $a + 1$; all the curves have $a + 1$ cusps. For the case $a = 3$, the curve is symmetrical about the y-axis.

Common features include: all the curves pass through $(0, -1)$; all the curves have a vertical asymptote at $x = -k$; all the curves have an oblique asymptote at $y = x - k$.

For the cases $k = 0.5$ and $k = 2$ the curves intersect the x-axis twice; once when x is positive and once when x is negative. These cases also have two stationary points.

(iii)

(ii)

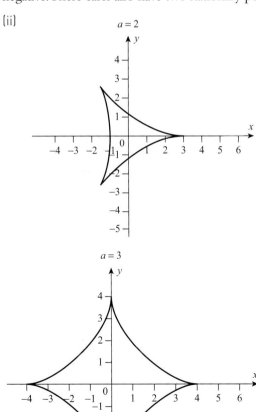

$a = 2$

$a = 3$

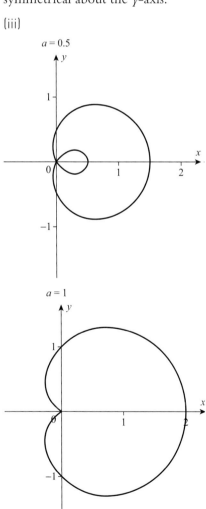

$a = 0.5$

$a = 1$

$a = 2$

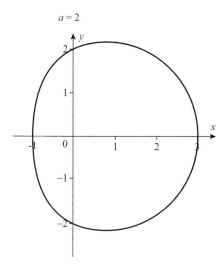

Common features include: all the curves pass through the point $(a + 1, 0)$; all the curves are symmetrical about the line $\theta = 0$; all the curves are bounded by a circle centred at the origin with radius $a + 1$.

For the cases $a = 0.5$ and $a = 1$, the curves pass through the origin/pole; the curve has a loop; for the case $a = 1$ the curve has a cusp.

2 (i)

$$\big((r\cos\theta + 1)^2 + (r\sin\theta)^2\big)\big((r\cos\theta - 1)^2 + (r\sin\theta)^2\big) = 1$$

$$(r^2 - 2r\cos\theta + 1)(r^2 + 2r\cos\theta + 1) = 1$$

$$r^4 - 4r^2\cos^2\theta + 2r^2 + 1 = 1$$

$$r^4 - 4r^2\cos^2\theta + 2r^2 = 0$$

Factorising gives: $r^2(r^2 - 4\cos^2\theta + 1) = 0$

$$r^2 - 4\cos^2\theta + 2 = 0 \Rightarrow r^2 = 2(2\cos^2\theta - 1)$$

$$r^2 = 2\cos 2\theta$$

(ii) $r = 2a\sin\theta\tan\theta$

$$r^2 = 2ar\sin\theta\tan\theta$$

$$x^2 + y^2 = 2ay\frac{y}{x}$$

$$y^2 = \frac{x^3}{(2a - x)}$$

(iii) $r = 2a\sin\theta\tan\theta$

$$x = 2a\sin t\tan t\cos t$$

$$= 2a\sin^2 t$$

$$y = 2a\sin t\tan t\sin t$$

$$= 2a\sin^2 t\tan t$$

(iv) $t = \dfrac{x}{a}$

$$y = \frac{a}{1 + \left(\dfrac{x}{a}\right)^2}$$

$$y = \frac{a^3}{a^2 + x^2}$$

Discussion point (Page 8, upper)

$-2\sin 2t = 0 \Rightarrow t = 0, \dfrac{\pi}{2}, \pi, \dfrac{3\pi}{2}$; however, the

numerator $\cos t$ is also 0 at $t = \dfrac{\pi}{2}$ and $t = \dfrac{3\pi}{2}$. The gradient of the tangent will be parallel to the y-axis when $t = 0$ and $t = \pi$ as the denominator of the derivative is zero but the numerator is non-zero. The gradient of the tangent to the curve at $t = \dfrac{\pi}{2}, \dfrac{3\pi}{2}$ is not defined.

Discussion point (Page 8, lower)

$\dfrac{dr}{d\theta}$ represents the rate of change of the distance from the pole with respect to the angle. It can be thought as comparing it to the rate of increase of r on the spiral $r = \theta$, which has derivative $\dfrac{dr}{d\theta} = 1$.

Exercise 1.2 (Page 10)

1 (i)

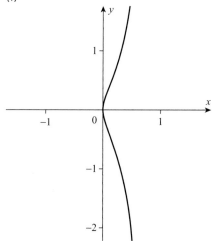

(ii) $\dfrac{dy}{dx} = \dfrac{3t^2 + 1}{\left(\dfrac{-2t^3 + 2t(t^2 + 1)}{(t^2 + 1)^2}\right)}$

$$= \frac{3t^6 + 7t^4 + 5t^2 + 1}{2t}$$

As the numerator is the positive sum of even powers of t, plus 1, it is ≥ 1 and therefore $\dfrac{\mathrm{d}y}{\mathrm{d}x}$ is never 0, and so the tangent to the curve is never parallel to the x-axis. The tangent is parallel to the y-axis when $t = 0$, which is the point $(0, 0)$.

2 The gradient of the tangent is $\dfrac{\mathrm{d}y}{\mathrm{d}x} = -\dfrac{1}{x^2}$.

The equation of the tangent at the point $x = x_1$ is $y - \dfrac{1}{x_1} = -\dfrac{1}{x_1^2}(x - x_1)$ or $y = \dfrac{2x_1 - x}{x_1^2}$. Solving $\dfrac{1}{x} = \dfrac{2x_1 - x}{x_1^2}$ gives $(x - x_1)^2 = 0$, i.e. the only point where the tangent intersects the curve is at $x = x_1$ (the point at which the tangent is drawn).

Exercise 1.3 (Page 14)

1 (i)

$k = -1$

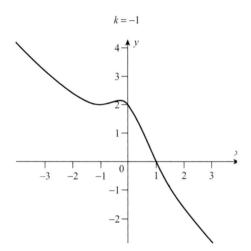

$k = 0$

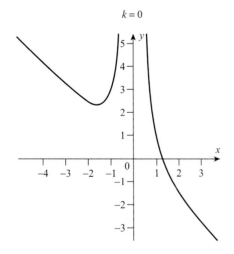

$k = 4$

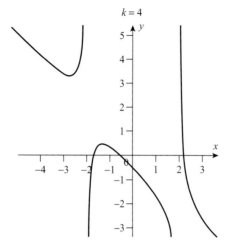

(ii) The equation can be written $y = -x + \dfrac{2}{x^2 - k}$.

(a) $k > 0$: Oblique asymptote $y = -x$, vertical asymptotes $x = -\sqrt{k}$ and $x = \sqrt{k}$.

(b) $k = 0$: Oblique asymptote $y = -x$, vertical asymptote $x = 0$.

(c) $k < 0$: Oblique asymptote $y = -x$.

2 (i)

$k = -1$

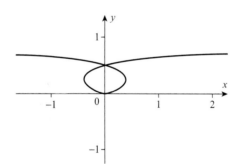

$k = 0$

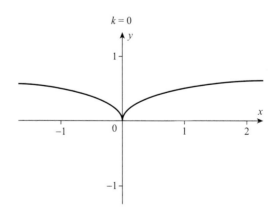

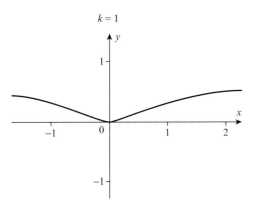

$k = 1$

(ii) For the case $k = 0$:

The cusp is at $(0, 0)$. Substituting $t = 0$ into the parametric equations gives $x = 0$ and $y = 0$; therefore the point is defined.

$$\frac{dy}{dx} = \frac{\left(\dfrac{-2t^3 + 2t(t^2 + 1)}{(t^2 + 1)^2}\right)}{3t^2}$$

$$= \frac{2}{3t^5 + 6t^3 + 3t}$$

As t tends to 0 from below, $\dfrac{dy}{dx} \to -\infty$.

As t tends to 0 from above, $\dfrac{dy}{dx} \to \infty$.

The tangent to the curve tends to vertical as the point is approached from either side, and hence the point is a cusp.

Exercise 1.4 (Page 20)

1 $2y^3 + 2x^2y + x^2 = 0$

2 (i)

$$\int_0^a \sqrt{(2x)^2 + 1}\, dx = \frac{2a\sqrt{4a^2 + 1} - \ln(\sqrt{4a^2 + 1} - 2a)}{4}$$

(ii) $\displaystyle\int_0^1 \sqrt{\left(-2\frac{t}{(t^2 + 1)^2}\right)^2 + \left(\frac{1 - t^2}{(t^2 + 1)^2}\right)^2}\, dt = \frac{\pi}{4}$

(iii) $\displaystyle\int_0^{2\pi} \sqrt{(-a\sin\theta)^2 + (a(\cos\theta + 1))^2}\, d\theta = 8a$

Chapter 2

Discussion point (Page 25)

Isoclines connect points on the family of particular solutions such that the tangents have the same gradient, and, in general, this will not be the curve for a particular solution. However, if there are particular solutions that are straight lines, these will have the same gradient along the particular solution and therefore will also be isoclines.

For example, the particular solution to the differential equation $\dfrac{dy}{dx} = y - 2x$ that passes through $(0, -2)$ is $y = 2x + 2$, which is also an isocline featuring all the points where the gradient of the tangent is 2. Similarly, if a differential equation has a particular solution of the form $y = k$, then this is also an isocline connecting all the points with gradient 0.

Activity 2.1 (Page 25)

You could try taking a differential equation you are familiar with, such as $\dfrac{dy}{dx} = x^2$, and add a parameter to obtain the differential equation $\dfrac{dy}{dx} = x^2 + k$. Plot the tangent field for different values of the parameter and explain how these relate to the solutions to the differential equation.

Exercise 2.1 (Page 25)

1 (i) (a) $a = 1$

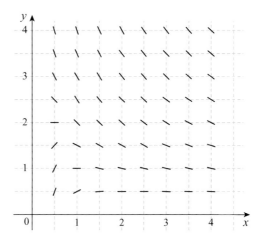

(b) $a = 4$

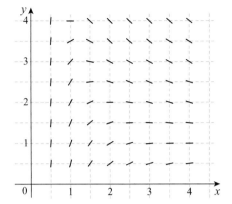

(ii) $\dfrac{a - xy}{x^2} < 0 \Rightarrow a < xy$, i.e. $y > \dfrac{a}{x}$.

2 (i)

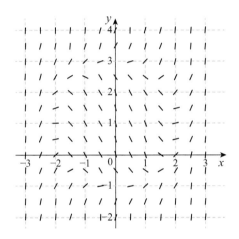

(ii) $x^2 + y^2 - 2y - 3 = k$ can be written
$x^2 + (y - 1)^2 = k + 4$; therefore isoclines are
circles centred on $(0, 1)$.

Activity 2.2 (Page 27)

You could use some differential equations from
MEI A Level Mathematics Year 2 or *MEI A Level Further
Mathematics Year 2*. These books will present the pen
and paper method for solving them but you can try
solving them with CAS, verifying the solution and
then plotting the resulting curve with a parameter in
place of the constant of integration.

Exercise 2.2 (Page 27)

1 For the differential equation $\dfrac{dy}{dx} = 2y - x$:

(ii) $y = c\,e^{2x} + \dfrac{x}{2} + \dfrac{1}{4}$

(ii) $\dfrac{dy}{dx} = 2c\,e^{2x} + \dfrac{1}{2}$

$= 2\left(c\,e^{2x} + \dfrac{x}{2} + \dfrac{1}{4}\right) - x$

$= 2y - x$

(iii) The three distinct cases are given by $c<0$, $c=0$,
and $c>0$, e.g. $c=-1$, $c=0$ and $c=1$.

$c = -1$

$c = 0$

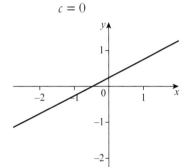

$c = 1$

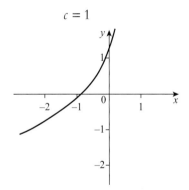

All cases have an oblique asymptote at $y = \dfrac{x}{2} + \dfrac{1}{4}$ as
$x \to \infty$.

$c = -1$ has a turning point that is a maximum.
$c = 0$ is a straight line.
$c = 0$ and $c = 1$ are increasing functions.

2 (i) $y = \dfrac{2}{2c - x^2}$

(ii) $\dfrac{dy}{dx} = \dfrac{4x}{(2c - x^2)^2}$

$= \left(\dfrac{2}{(2c - x^2)}\right)^2 x$

$= x^2 y$

(iii) The three distinct cases are given by $c<0$,
$c=0$ and $c>0$, e.g. $c = -1$ $c = 0$ $c = 1$

$c = -1$

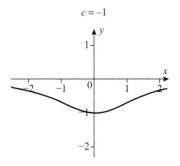

$c = 0$

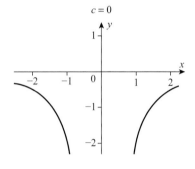

$c = 1$

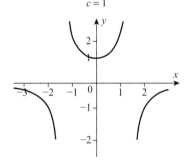

All cases have a horizontal asymptote at $y = 0$.

None of the cases cross the x-axis.

$c = -1$ and $c = 1$ have a turning point at $\left(0, \frac{1}{c}\right)$.

$c = 0$ has a vertical asymptote at $x = 0$.

$c = 1$ has vertical asymptotes at $x = \pm\sqrt{2c}$.

Activity 2.3 (Page 31)

Using step sizes of $h = 0.0125$ and $h = 0.00625$ give approximations that agree to 2 decimal places: $y = 1.36$.

Activity 2.4 (Page 33)

Using step sizes of $h = 0.0125$ and $h = 0.00625$ give approximations that agree to 3 decimal places: $y = 1.362$.

Activity 2.5 (Page 35)

Using step sizes of $h = 0.025$ and $h = 0.0125$ give approximations that agree to 5 decimal places: $y = 1.36215$.

Exercise 2.3 (Page 35)

1 (i) The value for h is entered into cell A2. Columns for x, Y, k_1 and k_2 are created as follows:

B2 = 1, B3=B2+A$2
C2 = 3, C3=C2+(D2+3*E2)/4
D2 = A$2*(SQRT(C2)-B2/SQRT(C2))
E2 = A$2*(SQRT(C2+2*D2/3)-
(B2+2*A$2/3)/(SQRT(C2+2*D2/3)))

and the formulae are filled down.

(ii)
   (a) $y = 4.09319$
   (b) $y = 4.09312$
   (c) $y = 4.09310$.

(iii) Result looks secure to 3 d.p.: $y = 4.093$ and is likely to be secure to 4 d.p.: $y = 4.0931$.

2 (i) The value for h is entered into cell A2. Columns for x, Y, k_1, k_2, k_3 and k_4 are created as follows:

B2=0, B3=B2+A$2
C2=1, C3=C2+(D2+2*E2+2*F2+G2)/6
D2=A$2*(SIN(B2)+COS(C2))
E2=A$2*(SIN(B2+A$2/2)+COS(C2+D2/2))
F2=A$2*(SIN(B2+A$2/2)+COS(C2+E2/2))
G2=A$2*(SIN(B2+A$2)+COS(C2+F2))

and the formulae are filled down.

(ii)
   (a) $y = 1.69221441$
   (b) $y = 1.69221445$
   (c) $y = 1.69221446$

(iii) Result looks secure to 6 dp: $y = 1.692214$ and is likely to be secure to 7 dp: $y = 1.6922145$.

Chapter 3

Exercise 3.1 (Page 40)

1 (i)
```
def prog1(n):
    for i in range(1,n+1):
        if i%13==1 and i%15==2:
            print(i)
```
(ii) 92, 287, 482

(iii) Adapt the if statement to: `if i%12==1 and i%15==2`: Program returns no results.

(iv) If n is 1 more than a multiple of 12, then $n = 12a + 1$ or $n = 3(4a) + 1$. Likewise $n = 15b + 2$ or $n = 3(5b) + 2$. An integer cannot be both 1 more than a multiple of 3 and 2 more than a multiple of 3; therefore there are no such integers.

2 (i)
```
import math
def prog2(n):
    for i in range(1,n+1):
        t=i*(i+1)/2
        if math.floor
            (t**(1/2))==t**
            (1/2):
                print(t)
```

(ii) $1, 36, 1225, 41\,616$.

(iii) As T_n is square, $T_n = m^2$ for some m.

$$T_{4n(n+1)} = \frac{4n(n+1)(4n(n+1)+1)}{2}$$
$$= \frac{n(n+1)}{2}(16n^2+16n+4)$$
$$= m^2(4n+2)^2$$
$$= (m(4n+2))^2$$

Therefore as each value that is both a triangle and square number can be used to generate a larger example, and some examples exist, there must be infinitely many.

Discussion point (Page 41, upper)

Any factor of n must appear in a factor pair where $ab = n$. One of a or b must be $\leq \sqrt{n}$ (as, if not, $ab > n$). Therefore if there are no factors from 2 up to the largest integer $\leq \sqrt{n}$, then n has no factors other than 1 and itself.

Activity 3.1 (Page 41)

One approach you could take is to check if the number is divisible by 2 and then, if not, only check the odd numbers.

Discussion point (Page 41, lower)

This has been considered for the special cases $p_1 = 2$ and $p_1 = 3$ but is similar for other primes.

If $p_1 = 2$, then p_1p_2 is even and cannot be written as the product of two further primes, as q_1q_2 is odd.

If $p_1 = 3$, then p_1p_2 is a multiple of 3. Given two other primes q_1 and q_2, each of these will be either 1 more than a multiple of 3 or 2 more than a multiple of 3. The product of any two such numbers is never a multiple of 3 and therefore cannot be equal to p_1p_2.

Activity 3.2 (Page 42)

The Euclidian algorithm is usually more efficient than the example given in the text. They can be compared by finding the GCD of two large numbers.

Exercise 3.2 (Page 44)

1 (i) Using `primetest(n)` as defined in Section 3.2:
```
def prog3(n):
    for k in range(1,n+1):
        if primetest(2**k-1)==1:
            print(k,2**k-1)
```

(ii) $2, 3$
$3, 7$
$5, 31$
$7, 127$
$13, 8191$

(iii) If $2^k - 1$ is prime, then $2^{k-1}(2^k - 1)$ has factors
$1, 2, 4, ..., 2^{k-1}, (2^k - 1), 2(2^k - 1), 4(2^k - 1), ...,$
$2^{k-1}(2^k - 1)$.

The sum of the factors excluding $2^{k-1}(2^k - 1)$ is
$$2^k - 1 + (1 + 2 + 4 + ... + 2^{k-2})(2^k - 1).$$
$$= (2^k - 1) + (2^{k-1} - 1)(2^k - 1)$$
$$= 2^{k-1}(2^k - 1)$$

(iv) $6, 28, 496, 8128, 33\,550\,336$

2 (i) mn is a common multiple of m and n, and therefore the lowest common multiple of m and n cannot be greater than this.

(ii)
```
def prog4(m,n):
    for i in range(1,m+1):
        if (i*n)%m==0:
            return(i*n)
```

(iii) $108\,820\,808$

(iv) If m and n are not co-prime, then $m = ga$ and $n = gb$, where $g > 1$. gab is a multiple of both m and n and, as $mn = g^2ab$, then $gab < mn$.

Activity 3.3 (Page 46)

$(4k)^2 = 16k^2$
$\quad = 4(4k^2)$

$(4k+1)^2 = 16k^2 + 8k + 1$
$\quad = 4(4k^2 + 2k) + 1$

$(4k+2)^2 = 16k^2 + 16k + 4$
$\quad = 4(4k^2 + 4k + 1)$

$$(4k+3)^2 = 16k^2 + 24k + 9$$
$$= 4(4k^2 + 6k + 2) + 1$$

Discussion point (Page 46)

$_pC_n = \dfrac{p!}{n!(p-n)!}$. Both $n!$ and $(p-n)!$ are products of positive integers that are co-prime with p.

Discussion point (Page 47)

$a^p \equiv a \pmod{p}$ can be written $a^p - a = kp \Rightarrow a(a^{p-1} - 1) = kp$. If p is not a factor of a then p must be a factor of $a^{p-1} - 1$, giving $a^{p-1} \equiv 1 \pmod{p}$.

Extension activity (Page 47)

The Carmichael numbers n, where $n < 10\,000$, are 561, 1105, 1729, 2465, 2821, 6601 and 8911. They are all products of at least three distinct prime factors.

Activity 3.4 (Page 48)

For example, 3 and 20 are co-prime and $\varphi(20) = 8$. The eight values m ($m < 20$), where m and 20 are co-prime are $m = 1, 3, 7, 9, 11, 13, 15, 17$ and 19. Considering $3m$ modulo 20 for each value: $3 \times 1 \equiv 3$, $3 \times 3 \equiv 9$, $3 \times 7 \equiv 1$, $3 \times 9 \equiv 7$, $3 \times 11 \equiv 13$, $3 \times 13 \equiv 19$, $3 \times 17 \equiv 11$ and $3 \times 19 \equiv 17$. This gives $3^8 \times 1 \times 3 \times 7 \times 9 \times 11 \times 13 \times 17 \times 19 \equiv 1 \times 3 \times 7 \times 9 \times 11 \times 13 \times 17 \times 19 \pmod{20}$, which can be simplified to $3^8 \equiv 1 \pmod{20}$.

Discussion point (Page 48)

Yes. Show that each value of $am \pmod{n}$ must also be co-prime with n and that no two values can be the same.

Activity 3.5 (Page 49)

For $p = 11$: $2 \times 3 \times 4 \times 5 \times 6 \times 7 \times 8 \times 9 = (2 \times 6) \times (3 \times 4) \times (5 \times 9) \times (7 \times 8)$, each pair of which is congruent to 1 modulo 11.

For any value of p, each value of a, where $2 \leq a \leq p - 2$, will have a value b, where $2 \leq b \leq p - 2$, such that $ab \equiv 1 \pmod{p}$ and $a \neq b$.

Discussion point (Page 49)

If n is composite and not the square of a prime number, then $(n-1)!$ will be a product of positive

integers that contain at least one factor pair of n. Therefore $(n-1)!$ is a multiple of n and $(n-1)! \equiv 0 \pmod{n}$.

If n is composite and the square of a prime number p, where $p > 2$, then both p and $2p$ appear in $(n-1)!$ and hence $(n-1)!$ is a multiple of p^2 and $(n-1)! \equiv 0 \pmod{n}$. If $p = 2$, then $n = 4$ and $(n-1)! \equiv 2 \pmod{n}$.

Exercise 3.3 (Page 49)

1 (i)
```
def prog5(m,n):
    for a in range(0,n):
        for b in range(0,n):
            if ((a**n)+(b**n))%n==m:
                print(a,b)
```

(ii) (a) $(0,5), (1,4), (2,3), (3,2), (4,1), (5,0), (6,6)$.

(b) $(0,0), (1,12), (2,11), (3,10), (4,9), (5,8), (6,7), (7,6), (8,5), (9,4), (10,3), (11,2), (12,1)$.

(iii) If n is prime, then $a^n \equiv a \pmod{n}$ and $b^n \equiv b \pmod{n}$ by Fermat's last theorem. Hence $a^n + b^n \equiv m \pmod{n} \Rightarrow a + b \equiv m \pmod{n}$.

2 (i) Using `gcd(m,n)` as defined in Section 3.2:
```
def prog6(n):
    for i in range(1,n):
        if gcd(i,n)==1:
            print(i)
```

(ii)
(a) 1, 3, 5 and 7 are co-prime with 8.

(b) 1, 2, 3, 4, 6, 7, 8, 9, 11, 12, 13, 14, 16, 17, 18, 19, 21, 22, 23 and 24 are co-prime with 25.

(iii) Positive integers co-prime with p^a are:
$1, 2, \ldots, (p-1)$

$p + 1, p + 2, \ldots, (2p - 1) \cdots$

$p^a - p + 1, p^a - p + 2, \ldots, (2p - 1)$

i.e. p^{a-1} sets of $p - 1$ integers.

$$\varphi(p^a) = p^{a-1}(p-1)$$
$$\varphi(8) = \varphi(2^3)$$
$$= 2^{3-1}(2-1)$$
$$= 4$$
$$\varphi(25) = \varphi(5^2)$$
$$= 5^{2-1}(5-1)$$
$$= 20$$

Activity 3.6 (Page 52)

For example, the first few solutions to $x^2 - 3y^2 = 1$ are $x = 2, y = 1; x = 7, y = 4; x = 26, y = 15;$ $x = 97, y = 56.$

$(2 + \sqrt{3} \times 1)^2 (2 - \sqrt{3} \times 1)^2 = (7 + \sqrt{3} \times 4)(7 - \sqrt{3} \times 4);$

$(2 + \sqrt{3} \times 1)^3 (2 - \sqrt{3} \times 1)^3 = (26 + \sqrt{3} \times 15)(26 - \sqrt{3} \times 15);$

$(2 + \sqrt{3} \times 1)^4 (2 - \sqrt{3} \times 1)^4 = (97 + \sqrt{3} \times 56)(97 - \sqrt{3} \times 56).$

Discussion point (Page 52, upper)

Yes. This can be shown for the smallest solution x_1, y_1 and a larger solution x_k, y_k: repeated division of $x_k + \sqrt{n}y_k$ must lead to $x_1 + \sqrt{n}y_1$.

Discussion point (Page 52, lower)

No. If n is a square number, then

n can be written m^2, giving

$x^2 - m^2 y^2 = 1$, i.e. $(x + my)(x - my) = 1$. The only

integer solutions to this are $(x + my) = (x - my) = 1$

or $(x + my) = (x - my) = -1$, both of which require

$m = 0$ or $y = 0$.

Activity 3.7 (Page 53)

Some small solutions to $a^3 + b^3 = c^3 + 1$ are:
$a = 9, b = 10, c = 12$; $a = 64, b = 94, c = 103$;
$a = 73, b = 144, c = 150.$

Exercise 3.4 (Page 54)

1   (i) Using `primetest(n)` as defined in Section 3.2:

```
def prog7(n):
 for c in range(1,n):
  for b in range(1,c):
   for a in range(1,c):
    if a*a+b*b==c*c and primetest(a)==1:
     print(a,b,c)
```

(ii) $(3, 4, 5), (5, 12, 13), (7, 24, 25), (11, 60, 61),$ $(13, 84, 85).$

(iii) $2^2 = c^2 - b^2 \Rightarrow 4 = (c + b)(c - b).$
This requires either $(c + b) = 4, (c - b) = 1$ or $(c + b) = 2, (c - b) = 2$. Neither of these have positive integer solutions for c and b. Therefore $a = 2$ is not possible in any solution to $a^2 + b^2 = c^2.$

(iv) By symmetry, $b = 2$ is not possible in $a^2 + b^2 = c^2$. Any solution with a and b prime would require them both to be odd numbers. As odd^2 + odd^2 = odd + odd, and hence c^2 is even $\Rightarrow$ is even. As $c = 2$ is not possible, then c cannot be prime.

2   (i)
```
def prog8(m,n):
    for x in range(1,m):
        for y in range(1,m):
            if x*x-n*y*y==-1:
                print(x,y)
```

(ii) (a) $x = 1, y = 1; x = 7, y = 5; x = 41, y = 29.$
(b) $x^2 - 4y^2 = -1$ has no solutions.
(c) $x = 2, y = 1; x = 38, y = 17.$

(iii) If $n \equiv 0 \pmod 4$, then $x^2 - ny^2 = -1$ is equivalent to $x^2 - 0 \equiv 3 \pmod 4$, i.e. $x^2 \equiv 3 \pmod 4$. $x^2 \equiv 3 \pmod 4$ is not possible as for all positive integers n, $n^2 \equiv 0 \pmod 4$ or $n^2 \equiv 1 \pmod 4$, and therefore this has no solutions.